RADAR

RADAR

FOR MARINE
NAVIGATION & SAFETY

JACK WEST

 VAN NOSTRAND REINHOLD COMPANY
New York Cincinnati Toronto London Melbourne

Published in 1978 by Van Nostrand Reinhold Company
A division of Litton Educational Publishing, Inc.
135 West 50th Street, New York, NY 10020,

Van Nostrand Reinhold Limited
1410 Birchmount Road, Scarborough,
Ontario M1P 2E7, Canada

Van Nostrand Reinhold Australia Pty. Limited
17 Queen Street, Mitcham, Victoria 3132, Australia

Van Nostrand Reinhold Company Limited
Molly Millars Lane, Wokingham, Berkshire, England

16 15 14 13 12 11 10 9 8 7 6 5 4 3 2 1

Library of Congress Cataloging in Publication Data

West, Jack.
    Radar for marine navigation and safety.

    Bibliography: p.
    Includes index.
    1. Boats and boating—Radar equipment.   2. Radar in
navigation.   3. Collisions at sea—Prevention.   I. Title.
VM325.W47   623.89′33   72-12480
ISBN 0-442-29353-4

CONTENTS

INTRODUCTION

In the late 1950s my wife and I were first exposed to radar. What stimulated our interest was a radiotelephone conversation with a friend who then had radar on his cruiser. We were approaching each other in moderate fog and knew each other's approximate position. My friend commented over the radiotelephone that we were about to overtake a small boat on our starboard side. Within a few minutes we did, as it loomed through the fog about a hundred yards away.

Shortly after that we decided it was time that we had radar, particularly because we planned some extensive coastal cruising. Since then radar has been a trusted shipmate and crew member aboard our boats. We have safely found our way into harbors on both the Pacific and Atlantic coasts that we would not have dared to enter without radar; we have assisted in rescues at sea that could not have been accomplished without radar's ability to see through the thickest fogs; over the years we have proven to ourselves and many others that radar is no longer a status symbol but an important and reliable navigational aid.

In contrast to insurance which pays only after-the-fact, radar pays off every minute it is in use. We have said many times that prevention of an accident or a close-call is far better than all the insurance in the world—and radar can do just that, if properly used.

The purpose of this book is to bring into focus the fundamentals of radar, what it can and cannot do, how to install it, and how to properly use it. I sincerely hope that it will contribute to greater safety of boat operation and navigation, and that those reading the book will share with me the justifiable enthusiasm that I have for radar as an aid to the modern mariner.

Rancho Palos Verdes, Cal.

Jack West

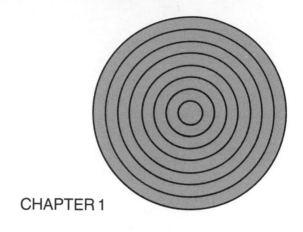

CHAPTER 1

WHY A RADAR?

To fully appreciate the answer to this question it is necessary to draw comparisons with some of the navigational tools with which we are all familiar. Every boat has a whistle, many boats have a depth sounder, and most boats are equipped with a searchlight. All three of these add safety to boat operation in the same fundamental manner that does a radar, and their basic principles of operation are identical. The principle is one of *reflecting* sound, light, or radio waves. Let's look at these safety devices more carefully.

The Whistle

For years the skippers of ships and small craft have used the whistle as a navigational aid during fog conditions. This is in contrast to use of the whistle for signalling purposes. The art of determining distance and bearing from landmarks was developed to its highest peak and usefulness in British Columbian and Alaskan waters. There tugs, fish boats, commercial ships, and yachts would "toot their way" from one island or headland to another. The same system was used on many rivers, where echo boards were located at turns against which the sound waves from the boat's whistle would be bounced. The direction from which the echo came would tell the pilot his relative position to the board.

Sound travels at the rate of about 1100 feet per second. A short blast of the whistle, if aimed toward an island, headland, or echo board will be reflected as an echo, and by counting the seconds of time between the blast and the echo, distance can be determined. If the land is 3300 feet away, it requires about 3 seconds for the whistle's sound to reach it, and another 3 seconds for the echo to return to the ears of the pilot. The total time between blasting the whistle and hearing the echo would thus be 6 seconds. Since it had to travel twice the actual distance to the island, headland, or echo board, the signal went a total distance of about 6600 feet. Dividing that by 2 gives the answer to the pilot that the echo returned from a point that was about 3300 feet away. To a modest degree the bearing of the echo dead ahead, to port or starboard could be determined by the pilot. In this case the whistle is the transmitter and the pilot's ears the receiver.

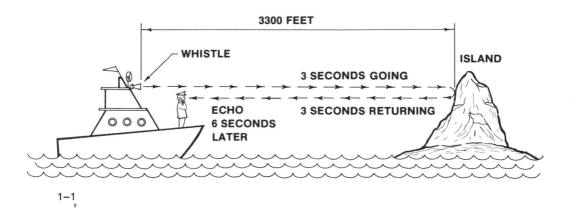

1–1

The Depth Sounder

The same principle of operation exists with depth sounders, except that the medium of travel is through water rather than air. Electrical impulses generated in the electronic circuitry of the sounder are converted to high frequency sound waves that are emitted from the transducer mounted on the boat's bottom. When an impulse leaves the transducer, which is both transmitter and receiver, it goes to the bottom and is bounced back. The time intervals are much shorter than with sound waves through air, however. In 1 second sound will go approximately 4800 feet through water, or about 2400 feet in 1/2 second. The depth of the water actually below the transducer is one-half these distances since the sound waves go down and back.

Most sounders on pleasure boats have a shorter range than that—such as from 120 to possibly 400 feet. With the latter, the time interval between emission of the signal and return of the echo is about 1/6 of a second. Presentation of the depth information on the recording, indicating, or digital type of sounder is merely a display of the fractions of a second required for the signal to go down and back, interpreted in feet or fathoms.

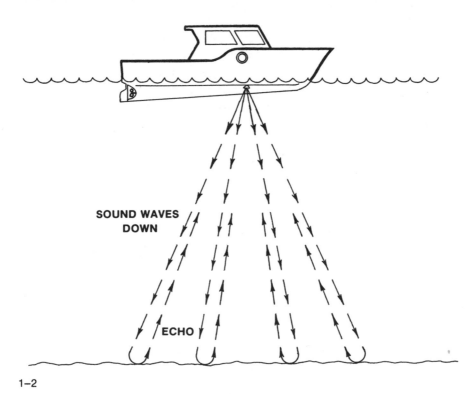

SOUND WAVES DOWN

ECHO

1–2

The Searchlight

With either the whistle or sounder, the distances are determined by measuring the seconds or fractions of a second of time for the signal to leave the boat and return as an echo. Through air or water the rate of travel of sound is relatively slow, compared with the rate that light travels through air. As noted earlier, in one second sound travels through air about 1100 feet, and through water about 4800 feet; but light travels about 162,000 nautical *miles* per second. This is also the same approximate speed of radio waves.

When a searchlight is rotated around the horizon at night, the beam of light must hit a buoy, cliff, breakwater, or other boat *within its range* to show anything. A small searchlight might have enough candlepower to illuminate an object at 1/4 mile—a more powerful one might be good for 1 mile. During the early stages of WW II, before the development of radar for aircraft defense purposes, many will remember seeing beams of tremendously powerful searchlights sweeping the skies. The beams would simply fade away to infinity unless they struck upon an airplane in flight or a cloud which would reflect the light waves to the eyes of the observers.

What occurs with the searchlight is directly comparable to the whistle or sounder, except at a vastly faster rate. Light waves emitted by the searchlight, when aimed at an airplane, buoy, or other object, are *reflected.* Obviously the speed of travel of light waves is so rapid that the time interval between emission and reflection is virtually instantaneous, and too short to be counted in seconds or small fractions.

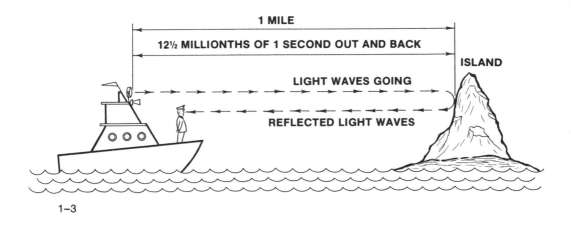

1 MILE

12½ MILLIONTHS OF 1 SECOND OUT AND BACK

ISLAND

LIGHT WAVES GOING

REFLECTED LIGHT WAVES

1–3

Of the three methods described—whistle, depth sounder, and searchlight—only the latter has a sufficiently narrow beam to provide a navigator with the kind of bearing information to steer a correct course. When the beam of the searchlight illuminates a buoy or other boat, its relative bearing can be precisely observed; even so, the searchlight is only a fair-weather aid, for it is useless in fog. With a whistle, only a moderate amount of bearing information is provided; and with a sounder, of the types used on pleasure boats, the bearing is merely *down.*

Early Radar Developments

It was not until the mid-1920s that the principles of today's radar were being recognized by scientists. To the British goes credit for finding that the height of the ionosphere could be measured by transmitting radio waves skyward and measuring the time before a reflected echo came back. In 1934 a shore-based radar they developed was able to detect aircraft in flight. Three years later, developments in the United States resulted in the installation of a crude radar device on the *U.S.S. Leary*—apparently the first naval vessel in the world to be radar equipped. In 1938 the British warships *Rodney* and *Sheffield* were also equipped. Ironically, the British battle cruiser *Hood*, which had been fitted with radar, was sunk by the Germans during WW II. A salvo from the *Hood* was directed at the *Bismarck* by an optical range finder rather than by its radar, and fell short of its target. The *Bismarck* returned fire, using their radar to direct its attack on the *Hood* with devastating accuracy.

The growing tensions in Europe during the late 1930s was a tremendous spur to the refinement of radar equipment by the scientists of Allied and enemy nations. In a short span of time radar was developed to a point that the periscopes of submarines could be detected miles away, and aircraft located far beyond the range of human sight. Military historians state that the early warning radar systems along the coasts of England made a decisive contribution in winning the Battle of Britain in 1940; later in the war they gave a substantial advantage to the Allies in all forms of military operations.

Use of the word *target* to refer to an object that reflected radio waves and showed as an echo on the scope was entirely correct during those war years. It was so indelibly stamped on the minds of radar observers, and those interested in radar, that its use has carried over to today's users of radar.

No other electronic device then, or even now, could pictorially display both bearing and range of targets, regardless of visibility. This is the principal answer to the question "Why A Radar?"

The original form of presentation, on what was termed an "A-scope," was soon supplanted with the Plan Position Indicator—generally referred to as PPI—type of radarscope. This presented a 360-degree view on the scope of all objects within the range to which the radar was adjusted showing distance and bearing to buoys, coastlines, and other craft both on the water and in the air. With this type of presentation the radar equipped boat is in the center of the scope, and what is shown is the same as what a chart would show within a circle corresponding to the range scale being used—plus the boats or other targets in the area.

After its military role, which accelerated its development and proved its value as a reliable device for *RA*dio *D*etection *A*nd *R*anging, radar for all forms of commercial vessel and aircraft usage was the logical step forward. By the late 1950s and early 1960s units became available for smaller craft, both commercial and pleasure. The development of transistors to reduce current consumption, reduction in physical size of components, increased rates of production that reduced costs—all have contributed to making radar a navigational aid that thousands upon thousands of boat owners now rely upon.

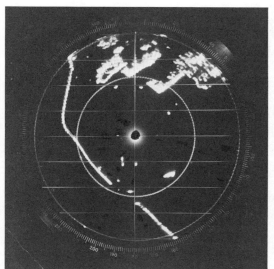 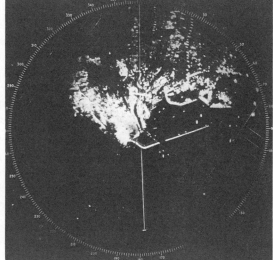

1–4a. Photograph of radarscope showing, at top over the bow, the entrance into Los Angeles harbor's main channel and astern the opening between the breakwaters. Radar was on the 1 mile range scale, with the ½ mile range ring indicating distance from breakwater opening astern and the main channel a few degrees to starboard off the bow. The Pilot Station is 5 degrees off the port bow on the ½ mile range ring. A large ship is shown off starboard stern quarter; other echoes are from small craft or channel buoys.

1–4b. Radarscope view of the approaches to Los Angeles harbor as seen from the Pilot Station where their shore-based radar is situated. The radar is on the 6 mile range scale.

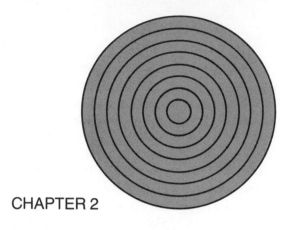

CHAPTER 2

PRINCIPAL USES OF RADAR

Too frequently radar is considered as a marine navigational aid too complicated for the average boat owner to understand or use, and, most frequently, only a device to help prevent collisions. Both assumptions are far from the truth. Radar is additionally used to track satellites, check speed of vehicles on highways, to show the pilot of an airplane his altitude above the terrain directly below him, and to detect approaching storms—to name only a few of the other uses.

Shipboard Radar

Two areas of usage apply to installations aboard boats: (1) as a navigational device when operating along coastal waters, in harbors, and in rivers to give the navigator accurate information as to his position; (2) and as an aid to prevent collisions with other craft.

For navigation, radar is of tremendous value—whether operating during clear weather or in the densest fog. This is primarily true because radar will indicate distance and bearing *simultaneously*. With one observation, the relative bearing and distance in miles to an identifiable landmark can be seen on the radarscope, and a fix established with a high degree of accuracy. In Fig. 2-1 a coastline is shown on the port side of the boat. The range scale is 1 mile, with the midcircle range ring repre-

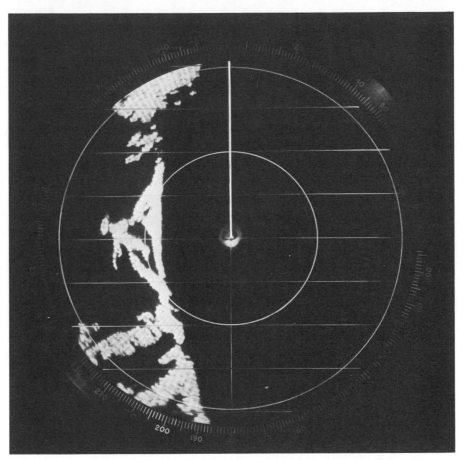

2-1

senting 1/2 mile. Two anchored boats are shown in the cove about 3/4 of a mile distant and 20 degrees to port. Distance of coastline off port beam is approximately 4/10 of a mile, and about 1 mile to the headland 10 degrees off the port bow.

In contrast to navigating with a radio direction finder, for example, radar-conspicuous landmarks exist along almost every mile of a shoreline, whereas radiobeacons are only located at certain places. Furthermore, it requires the presence of two radiobeacons in order to get bearings that can be crossed and plotted to establish a fix.

During clear weather, hand-bearing compasses or peloruses, are frequently used to establish a position, but just as with radio direction finders, two or more identifiable points must be in view as indicated in

Fig. 2-2. A single line-of-position from either an RDF, hand-bearing compass, or pelorus gives only the bearing to or from certain points but cannot tell the distance. And if visibility is poor, or no more than one radiobeacon is in the vicinity, the navigator must resort to dead reckoning.

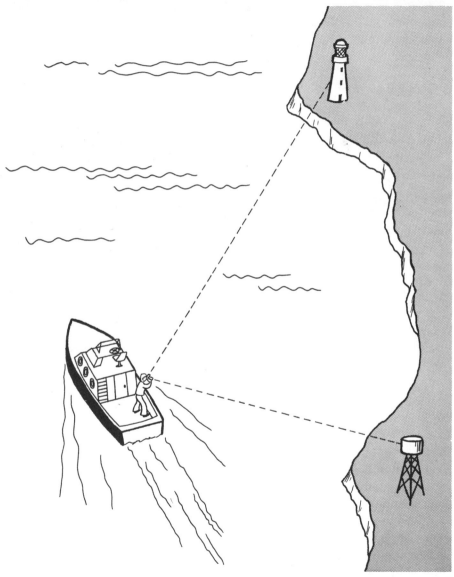

2-2

Prevention of collisions with other vessels is the second prime value of radar *when the radar is properly used.* Note that the statement is qualified, because radar by itself cannot assess the risk of collision, nor direct the craft to courses that should be taken to avoid a collision. Its role is solely to present information on the presence of other craft within the range of the radar. Unless the information is correctly interpreted by the navigator, and proper steps taken by him to avoid too close an approach to the other vessel, a dozen radars aboard one boat will not prevent a collision. Chapter 7 discusses the means of using radar information to avoid collisions, and Chapter 10 reviews some of the classic cases of collisions even though radar was aboard the vessels.

Shore-based Radar

As with the early development of shipboard equipment, so also have the British led in the use of shore-based radar for the guidance of vessels entering or leaving harbors and moving through rivers and to terminals. As long ago as 1947 the ferries plying the River Mersey between Liverpool and the opposite side of the river were given navigational assistance by radar when the river was blanketed in fog. The radar observer at the shore station used VHF radio to advise the ferry pilots of their positions as well as to alert them of other traffic in the near vicinity.

A year later, 1948, the first shore-based radar installation designed especially for harbor traffic supervision was also installed at Liverpool. The control station is located at the mouth of the River Mersey, and separate display consoles are used to show traffic within the ranges of radar scanners that cover the route from seaward as much as 20 miles, as well as various segments of the Liverpool Bay and up the river for an additional 4½ miles.

Many other harbor radar surveillance systems have been installed since the first one at Liverpool. Among them is London, which has eight separate radars spaced along the River Thames covering the entire area from its mouth to the docks in London. The displays are located at three control stations where the operators can observe all traffic and by VHF radiotelephone give information to the pilots as to their position and existence of other vessels traversing the river. Similar systems are on the Elbe River approaches to Hamburg, Germany. Worldwide, there are nearly 200 shore-based radars now providing assistance to marine traffic. Long Beach, Calif. has the first harbor surveillance installation made in the Western hemisphere, put into operation in 1949. Fig. 2-4 shows their harbor and approaches to it as displayed on a 16-inch diameter radarscope. It is significant that as the result of the system, shipping traffic into and from Long Beach is less adversely affected by severe fog than at the adjoining Los Angeles Harbor. At the latter, shore-based radar is not used for traffic control to the same extent as the one at Long Beach.

Harbor radar surveillance systems are similar to, but less sophisticated than, those for aircraft traffic control. Known originally as GCA—ground control approach—they have proven of inestimable value in smoothing and safeguarding the flow of aircraft traffic to and from airports. Aboard all transport aircraft, and many privately owned ones, a device termed a transponder is installed which emits an identifying signal when activated by the GCA radar. The identifying signal from the transponder is displayed on the controller's radarscope. In that manner the controller is able to identify each aircraft and can direct communications to the particular plane, giving the pilot information on other traffic near him, or vectoring the plane to a holding position; when other traffic is clear, the controller can direct the plane to the correct approach course to the airport and give the pilot final landing clearance.

2–3. Overlapping radar coverage of the Thames River from the Lower Reaches to the London docks is provided with seven short-range and one longer-range radars. A total of eleven displays are utilized at the three control stations. As ships travel upstream, traffic advisory information is first provided by the Garrison Point Station, then by Gravesend, and finally by the Thames House Station.

THAMES HOUSE

LONDON BRIDGE

☐ RADAR DISPLAY

GRAVESEND

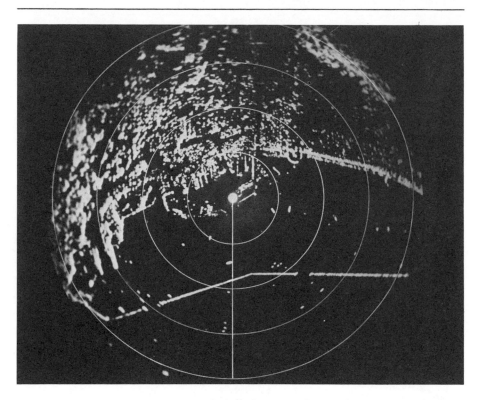

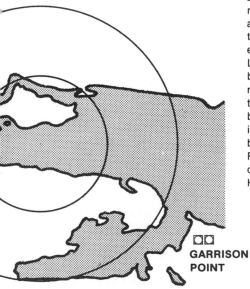

GARRISON POINT

2–4. Radarscope picture of Long Beach, California Harbor from their Pilot Station, taken shortly after installation of their harbor surveillance system. Top of illustration is north. Below and to either side of the center are the Long Beach and Los Angeles Harbors, with entrances through the breakwaters clearly shown. Range scale is 4 miles; each range ring is 1 mile apart. Since the time this picture was taken, the Pilot Station has been moved to a new location, a number of oil islands have been built to the east and Pier J has been constructed to the southeast of the Station. Refer to Fig. 1–4 in preceding chapter for a more detailed view of the entrance into Los Angeles Harbor's main channel.

Airborne Radar

Principal use of radar aboard civil aircraft is for detection of storm conditions that may lie in the flight path of the plane, but which cannot be seen by the pilot at night or while flying during times of low visibility. Flying at near sonic speeds requires bypassing of thunder clouds or other turbulent cloud formations to avoid severe strain on the aircraft and its passengers that would be felt when literally slamming into a storm cloud.

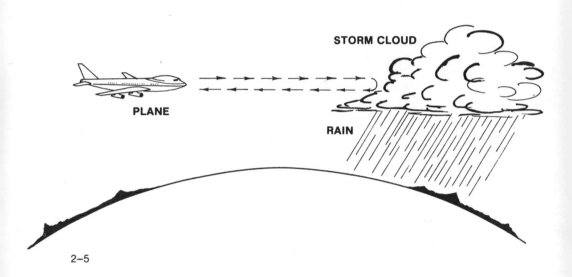

STORM CLOUD

PLANE

RAIN

2–5

In the nose sections of airliners is a parabolic antenna that sweeps back and forth through an arc of about 90 degrees horizontally. Storm clouds as much as 150 miles away show on the radarscope in varying degrees of brightness indicating areas of great or less cloud density. With this information the pilot can alter course to avoid the worst areas, or change altitude to fly over or below the clouds. As those who fly on commercial transports may recall, the "fasten seat belt" sign is often turned on when in smooth air—somewhat to the wonderment of the passengers. This is done by the pilot when his radar shows a storm front that could result in turbulent air possibly 100 miles ahead of him which would be reached within a few minutes.

The antennas of weather-radar aboard aircraft can also be tilted downwards to produce an oblique picture of the terrain below and ahead of the plane. Definition on the scope between water and land is very distinct, even at distances of 75 to 100 miles. On shorter ranges, particularly with radars designed for air-mapping, highways, major thoroughfares in cities, piers in harbors, bridges across rivers, and boats on harbor or ocean waters are clearly presented. This type of equipment requires an antenna system that rotates 360 degrees, as contrasted to the weather-radar type antennas. With this new generation of high-resolution radars, a 15-foot boat in heavy seas can be seen as far away as 10 miles from the searching plane or helicopter.

Radar for navigation by commercial aircraft is of less importance compared with their other electronic navigation devices. Omni/VOR beacons, distance measuring equipment, inertial navigation systems that even show the latitude and longitude of the aircraft at all times, are their primary navigational tools. But as one airline captain says, "It is sure nice to see the Golden Gate show up on radar when we are still 100 miles away and 40,000 feet above the fog shrouded Pacific, enroute from Honolulu to San Francisco."

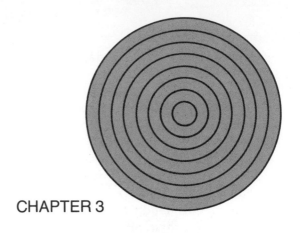

CHAPTER 3

PRESENT STATE OF THE ART

The principal difference between radar equipment available today and that which was used in the 1940s and 1950s lies in greater reliability, lower power consumption, more compact size, simpler controls for convenience of operation, and lower initial cost. There have been no basic changes in operating principles—a transmitter produces a signal; the signal is radiated by a rotating antenna (alternately called scanner or aerial); the returned echo comes back to the antenna; the signal is routed to the receiver where it is amplified; and finally, the signals from echoes are displayed on the cathode ray tube, or radarscope to show bearing and range.

Until the late 1950s, vacuum tubes were universally used in radars which drew considerable current and required periodic replacement; in the early 1960s, transistors reached a stage of development permitting their use by the more progressive radar manufacturers as a replacement to tubes; their use lowered current drain, increased reliability, and reduced the size of many sub-assemblies. The 1970s mark the use of even more advanced electronic developments such as integrated circuits, printed circuit boards, and miniaturization of many components—further increasing reliability and dependability.

The end result of these technological advances is equipment that can now provide the same high degree of reliability, safety, and assistance in navigation for the small craft operator that is available to professional navigators aboard merchant and naval vessels.

Types of Radar Sets

Broadly speaking there are two types: first, sets that place the transmitter/receiver (transceiver) and rotating antenna together, and second, those that have the transceiver separate from the antenna assembly. Both have the display unit in a separate single package along with the operating controls. The first type is generally referred to as a "two-unit" system comprising the antenna/transceiver assembly and the display unit; the other type is a "three-unit" system, since there is the antenna assembly, a separate transceiver, and the display.

Power supplies of many radar systems are built into the display unit or are integral with the antenna assembly, to convert the boat's power voltage to the higher voltages required by the radar. Some others, however, use external, or separate power supplies. In the preceeding discussion of two and three-unit systems any external power supply has not been counted as one of the units in the system.

"Two-unit" Systems

These systems have the transceiver and the open or radome-enclosed antenna together, with the display console mounted near the steering station. The primary advantage of these systems is that a flexible co-axial cable connects the transceiver/antenna assembly to the console. Generally the cable is about 3/4 inch in diameter, and can be routed down the supporting mast to the console with little difficulty; it can go through a pilot house roof or cabin top with only the need for a simple watertight fitting. The cable can be concealed in headliners and is not critical to the number of bends in it between the antenna assembly and the console. In some models two smaller diameter cables are used instead of a single one. A drawback to any two-unit system, that should be recognized, is that the antenna assembly (containing the transceiver) must be elevated above the deck to a height that may make it difficult to reach without a ladder for periodic servicing of the transceiver.

SCANNER

SCANNER

SCANNER TRANSCEIVER

WAVEGUIDE

CABLE

DISPLAY

DISPLAY

TRANSCEIVER

DISPLAY

3–1. These three Decca radars illustrate, at left, a "three-unit" system; in the center an open scanner "two-unit" system with the transceiver directly below the open antenna and a part of the antenna assembly; and at right a radome enclosed "two-unit" system. Waveguide connects the antenna and transceiver of the three-unit model; co-axial cables connect the antenna assembly to the display consoles of the open and radome "two-unit" systems.

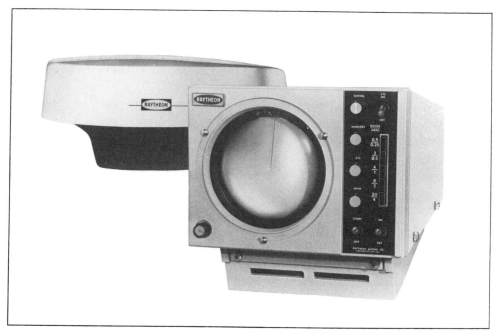

3–2. Raytheon model 2700 two-unit radar has a 5 kw peak output, five range scales of ½, 2, 4, 8, and 20 miles, 7 inch diameter scope, two pulse lengths of .08 and .4 microseconds, and a single pulse repetition frequency of 1500 per second. The 36 inch diameter radome-enclosed transceiver/antenna weighs 69 lbs. Indicator weight is 37½ lbs. Power drain is 156 watts.

3–3. The model X-10 radar imported from England by Ray Jefferson has four range scales of ½, 2, 5, and 10 miles, a 6 inch scope with 0–180 degree starboard and 0–180 degree port azimuth around it and engraved lines at each 10 degrees on the protective screen over the scope which are used for bearing determination. A variable electronic range ring is provided for distance measurement. Peak power is 3 kw; power drain is 36 watts. A single pulse length of .15 microseconds and a single repetition frequency of 750 per second results in a minimum range of 30 yards. Weight of radome-enclosed antenna and display unit is 34 pounds.

"Three-unit" Systems

As previously noted, these systems have an antenna assembly which consists only of the drive motor and the rotating scanner, a separate transceiver which is mounted below, and the display console. By placing the transceiver below, servicing of it is simpler. To accomplish this, however, generally requires the use of a waveguide between the antenna and the transceiver (instead of a flexible cable). Waveguide is a rectangular shaped hollow copper tube through which radio waves move to and from the transceiver and the antenna.

Waveguides are neither inexpensive nor simple to install. They are made to very exacting dimensional standards and are rigid and require precision fittings at each point where the waveguide changes direction, or where one section mates with another. Each mating point must be absolutely true and watertight. Besides the higher cost-per-foot of waveguide in comparison to the multistrand cable used with two-unit systems, the routing of the waveguide from antenna to transceiver requires more skill and care than when installing cable. High-power, long-range radars used on oceangoing vessels, freighters, passenger liners, and large yachts generally use the three-unit radar systems. This is because the antennas are frequently more than a 20-foot distant from the transceivers, which virtually dictates that waveguide be used because of the lower losses inherent with waveguide versus flexible cable.

Open or Enclosed Antennas

The trend in recent years for pleasure boat radar antenna systems is to enclose the scanner, drive motor, and the transceiver within a dome-like housing called a radome. There are advantages and disadvantages to enclosed antennas that require consideration.

In Chapter 5 the effect of antenna or scanner width on the overall performance of the radar is discussed in detail. Suffice it to say here, however, that the span or width of the scanner has a direct bearing on the beam width of radiated radio waves—the greater the span, the narrower the beam, which is desirable.

With open scanners, the span is usually considerably more than those enclosed within a radome. Unless an ungainly, large dome would be acceptable, there is a definite limitation on the span of the scanner when enclosed. Generally a dome of more than 35-inch diameter is not desirable on smaller craft.

Among the popular models of marine radars using enclosed scanners, the diameter of the domes will range from 30 to 35 inches which, aesthetically, is acceptable. The actual span of the scanner, however, is from 2 to 5 inches less than the diameter of the dome. This is in comparison with open scanners which generally range from 36 to as much as 72 inches in span, and because of the greater span they radiate a narrower beam which results in higher resolution of the radar picture.

The advantage of enclosed scanners, on the other hand, lies in three areas: reduced power to rotate the scanner since there is no wind resistance to its rotation, reduced risk of spray damaging the scanner and its drive mechanism, and particularly on sailboats, the elimination of the risk of fouling lines or sails.

Round or Rectangular Scopes

Historically the radarscopes have been round, ranging up to 16 inches or more in diameter. The pictorial presentation on the scope is generally with the boat in the center of the scope, and range of view equal in all directions for the full 360 degrees. Surrounding the scope is an azimuth graduated in degrees to obtain relative bearings on the echoes from other craft, buoys, or shorelines.

3-4. EPSCO/Brocks Seascan 3 kw, two-unit radar, has range scales of ½, 1, 2, 4, 8, and 16 miles displayed on a 6-inch diameter scope; two pulse lengths of .15 and .5 microseconds; and two pulse repetition frequencies of 2000 and 1000 times per second automatically selected by range in use. Radome-enclosed antenna has a 30-inch span. Weight of transceiver/antenna is 60 lbs.; weight of display is 13½ lbs.; power drain is 48 watts.

Some medium- and long-range types have means to offset the position of the boat from the center to increase the forward, aft, or side viewing ranges. There are other variations of presentation, such as true-motion, north-stabilized to give true north bearings, etc. that are used on naval and some ocean liners but these are usually beyond the budget or needs of the average pleasure boat owner.

For a time there were a number of radars available using rectangular scopes, although recently they have been withdrawn from the market by their manufacturers. The cathode ray tubes were generally higher than wide, with the boat's position below the center of the scope, providing forward range greater than aft or side-vision. Although their performance

3–5. Konel/Furuno model KRA-116 has five range scales of ½, 1½, 4, 8, and 16 miles, with 3 kw peak output. The antenna assembly consists of the transceiver and 36-inch span open scanner which rotates at 24 rpm. On the ½, 1½, and 4 mile ranges the pulse length is .08 microseconds; on the longer ranges it is .5 microseconds. Pulse repetition frequency is 1500 per second on all ranges. The 7-inch diameter scope is provided with a magnifier to give an apparent diameter of 12 inches. A front panel control permits off-centering of the display by as much as 25 percent of the range in use. Power drain is 90 watts. Antenna weight is 47 lbs.; display weight is 36 lbs.

was good, most boat owners wanted as much range abeam and astern as the forward range, particularly for coastal navigation. A rectangular scope radar rated, for example, at 16 miles forward range with 12 miles side vision and 13 miles astern, scans 696 square miles on each rotation of the antenna. This compares with 804 square miles by a round scope radar rated also at 16 miles.

Whether with a round or rectangular scope, the presentation is as seen by the navigator while looking over the bow, or at any other angle. If there is a buoy dead ahead, it will show on the scope as a white dot, or echo, dead ahead; a boat on the starboard beam shows as an echo at the same relative bearing on the scope. In addition, the radarscope will show the distance to the buoy or boat through use of the range rings that generally are electronically and automatically superimposed on the scope. Some lower priced radars have range rings printed or engraved on the face of the protective glass shield over the scope.

With either type scope, it is a cathode ray tube within which an electron gum emits a stream of electrons of varying strength. The stream is directed at the underside of the scope as a narrow beam which is visible as a rotating trace—synchronized with the rotation of the antenna. The underside of the tube's face is especially coated so that when a target returns an echo, the higher intensity of the electrons will create a bright spot. The coating material is such that the echo will continue to glow for a number of seconds, and it is reilluminated with the next rotation of the sweep.

Weights and Sizes

The principal components of a radar system are (1) the antenna which may or may not include the transceiver, and (2) the display console. Systems that utilize a separate transceiver, mounted below, will have a total weight for the three components somewhat more than the two-unit types of equipment.

Weight of the transceiver/antenna assemblies utilizing a radome will be less than those with an open antenna, mainly because the enclosed antenna will be smaller in span. Typically, one make with a 30-inch span scanner within a radome weighs 57 pounds. This is in comparison to 99 pounds for one that has an open scanner of 42-inch span. Both units are rated at 3 kw peak output.

Among the reasons for the greater weight of the open scanner systems is that a greater amount of power, and hence a larger driving motor is required to rotate the scanner. A Safety of Life at Sea conference requirement which applies to commercial ship radars and is sometimes followed by builders of pleasure craft radars, is that open scanners must be able to operate at their designed rotational speed in winds as high as 100 knots. Another factor which tends to increase weight is the need for more complex waterproof seals and bearings between the motor drive assembly and the antenna rotating shaft.

In considering the matter of antenna weight, the degree of optimum performance from the radar system becomes the principal factor. Although weighing more than radome enclosed scanners, the open type can be of substantially greater span. This will result in better performance. On the other hand, if weight topsides is a critical matter, the slight reduction in performance from a radome enclosed scanner is an acceptable compromise.

Weight and dimensions of the display console are fairly proportionate to the size of the "picture tube" radarscope and the maximum range of the radar. A 6-inch diameter scope of a 16 mile radar can be packaged within a console cabinet as small as 10 inches wide, 7 inches high and 12 inches deep. Total weight may be as little as 14 pounds. A 7-inch diameter scope of a 24 mile radar will require a slightly larger cabinet, and the weight may increase to around 25 pounds. For a 36-mile range radar, with a 9-inch diameter scope, the display console will measure about 20 inches wide, 12 inches high and 12 inches deep, and weigh about 48 pounds.

The combined weight of antenna system and display console of a recently introduced 10-mile range radar is as little as 34 pounds, a 16-mile range radar around 74 pounds, and about 150 pounds for a radar capable of 48 mile range.

Power Consumption and Range

Radars are rated as so-many kilowatts of peak output power, such as 1.5, 3, 5, 7 or more kw. Frequently these ratings are interpreted as meaning the amount of power required from the boat's electrical system. This is incorrect. Peak power is the amount of radiated power of the microsecond-length pulses of radio waves. Through the circuitry of the transmitter a few watts of power continuously used by the radar can be converted to much higher values when it is momentarily triggered for transmission. As will be discussed in Chapter 4, these short bursts or pulses of radio energy occur in millionths of one second, or microseconds. A typical 1000-microsecond on/off cycle will have an actual transmission period of only 1/2 microsecond, with 999.5 microseconds *off*; hence, a relatively low amount of input power from the boat's batteries can be converted to kilowatts of radiated power for infinitesimally short periods of time.

A radar with a maximum range of 48 miles, rated at 6 kw peak output, will draw up to 240 watts from the boat's electrical system. A smaller radar with a maximum range of 12 miles, rated at 3 kw, will draw around 85 watts. Neither amounts of power create a problem on modern boats with engines that are provided with alternators capable of delivering from 30 to 60 amps to keep the batteries charged.

In general, the range of a radar is related to its mean peak power output, receiver sensitivity, and antenna size. There are also other factors, however, that influence the maximum range. The most important is height of the target above the water and to a lesser extent the height of the antenna on the boat. High-frequency radio waves are basically line-of-sight plus about 15 percent which means that the antenna must be able to "see" the target. The table below shows the approximate number of nautical miles (n.m.) that radar will "see" based on antenna or target height, or a combination of both.

Height in ft.	Distance in n.m.	Height in ft.	Distance in n.m.	Height in ft.	Distance in n.m.
5	2½	55	9	110	13
10	4	60	9½	130	14
15	4½	65	10	150	15
20	5½	70	10	170	16
25	6	75	10½	190	17
30	6½	80	11	215	18
35	7	85	11	240	19
40	7½	90	11½	265	20
45	8	95	12	320	22
50	8½	100	12	380	24

Extreme elevation of a radar antenna is not particularly necessary. At 10 feet above the water, the radar horizon is 4 miles. Elevation of the antenna to 20 feet increases the range to the horizon to 5 1/2 miles—only a 40-percent increase in range with a 100-percent increase in antenna height. Whether the radar is rated at 3 kw or 100 kw, if the antenna is only 20 feet high, the horizon will still be only 5 1/2 miles away. To see a target at 15 miles, from a 20-foot high antenna, the target must be 70 feet high. This is derived by the 20-foot high antenna having a range of 5 1/2 miles,

and the 70-foot target having a visibility of 10 miles—adding the two together to get 15 1/2 miles. The principal benefit of higher radiated power is the ability to display a sharper and brighter picture on the radarscope from distant targets.

Other factors that influence the clarity of the picture from distant targets include the efficiency of the circuitry, the antenna span, sensitivity of the receiver, and composition of the target. These are more important than the amount of radiated power, particularly when it is realized that a sixteen-fold increase in power, assuming no other changes, only doubles the range of the radar.

A technological accomplishment in recent years has been an increase in the ranges of radars along with a decrease in peak power output. Whereas a few years ago a 7 1/2 kw radar had a maximum range of only 20 to 25 miles, modern radars with 3 kw are available with up to 48 mile range, due to improvements in circuit design and overall efficiency.

While it is true that long range radars have distinct advantages for offshore navigation, a trend in the opposite direction particularly useful to yachtsmen has been the inclusion of short range scales such as 1/4,

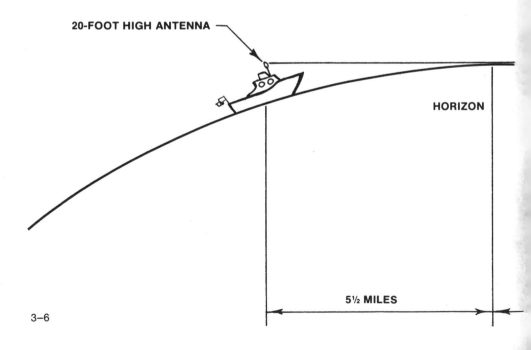

20-FOOT HIGH ANTENNA

HORIZON

5½ MILES

3–6

1/2, and 3/4 miles. And while some offshore operations may be beyond the range of a 12-mile radar but still within the range of an 18-mile unit, the capabilities of the shorter ranges, for close-in coastal operations, should not be overlooked. Sharp definition of targets on the short ranges may be critically important in dense fog and congested channels.

Recognition of the need for "radar eyes" focused specifically on short ranges has resulted in at least one manufacturer producing a unit with a maximum range of only 2500 yards (about 1 1/4 miles). It is designed solely for close-in operations, and not intended as a unit for coastal navigation. Such equipment fills an important need for many boatowners where weight and power consumption must be kept to low levels, yet in need of radar when approaching or operating in harbors during adverse weather or at night.

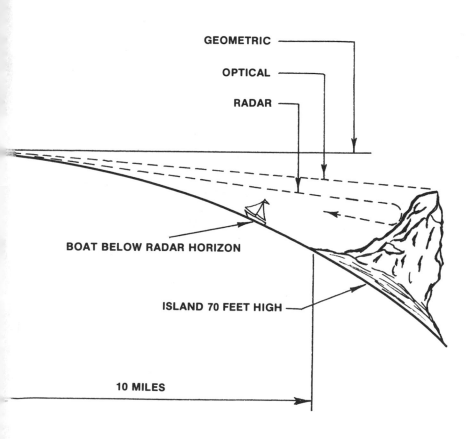

Cost of Equipment and Installation

There have been dramatic reductions in cost of radar systems in recent years. Excellent equipment, thoroughly proved in service, is available at prices between $1700 and $3750. They can also be leased, in some areas, by those who prefer not to make a capital investment. The physical installation has many variables, however, depending upon the location of the antenna and the display console. Many boats can accommodate the console without radical carpentry expense. There are some buyers, however, who may want to completely enclose the console within a cabinet when not in use, with provision to swing it out when needed. This can involve considerable expense. Another may elect to mount the console on an existing vertical or horizontal surface, which involves little expense.

3–7. The Bonzer Nautic-Eye short-range radar has four range scales of 100, 250, 1000, and 2500 yards displayed on a 5-inch diameter scope. Minimum range is 10 yards as the result of a short pulse length of .05 microseconds and a pulse repetition frequency of 33,000 per second. Transmitter frequency is in the 2900–3100 mHz band, whereas most other small craft radars operate in the 9415–9475 band. Unlike other radome-type radars where the fixed dome encloses a rotating scanner, the Nautic-Eye helical array antenna is a part of the radome which rotates at 20 rpm. Antenna/transceiver weighs 17 lbs.; display console weights 5¼ lbs.; power drain is 30 watts at 11–30 vdc.

Prefabricated antenna mounts are available for some models of radar. Generally they will be less expensive than having one fabricated by a local supplier, but still may range from $75 to $250 in cost, depending on their size and the type of installation.

Some boatowners have had enough experience to be able to physically mount the antenna assembly, the viewing console, transceiver, and power supply (if they are separate) in a matter of from 5 to 15 hours' time. The wiring together of the components, however, should be done by a qualified electronic technician, preferably from the firm that sold the radar. Furthermore, it is an FCC requirement that the first operation of the radar be done by an FCC radar-licensed technician.

3–8. Si-Tex/Koden model 22, two-unit radar, has 6 kw peak power and a maximum range of 48 miles. The seven range scales are ½, 1, 3, 6, 12, 24, and 48 miles; minimum range is 25 yards. Dual pulse lengths of .07 and .13 microseconds and dual repetition frequencies of 1650 and 550 per second are automatically switched when changing from short to long ranges. Open scanners of 3, 4, and 6-foot span are optional and weigh from 81½ to 100 lbs.; display console weight is 60 lbs. Power drain ranges between 160 and 240 watts depending upon voltage.

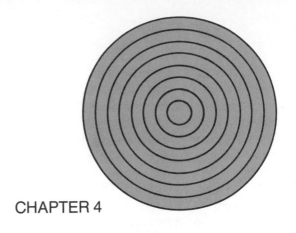

CHAPTER 4

RADAR PRINCIPLES AND COMPONENTS

An extension of the comparison of radar with a depth sounder, whistle, or searchlight can also be made with a radiotelephone. In the latter, a transmitter generates radio waves that carry the voice communication and a receiver hears the transmissions from another boat or shore station. When the microphone button is pressed, a relay connects the antenna to the transmitter section of the radiotelephone and disconnects the receiver; when the button is released the relay reconnects the antenna to the receiver section. This is a mechanical means of switching—as compared with electronic switching used in a radar to isolate the receiver during the time that radio waves are emitted from the transmitter, and visa versa. With the radiotelephone this switching cycle is between transmitting and receiving which may be only once or twice per *minute;* but with radar, the switching must be done from 500 to 3000 times per *second*, depending upon the pulse repetition frequency of the equipment.

A further distinction should be made at this point between radiotelephones and radars. Radiotelephone waves are domelike nondirectional. They radiate in a full 360-degree pattern from the antenna on a horizontal path; they also go skyward. With radar, however, the radiated radio waves are concentrated and beamed in a narrow path horizontally and vertically. The horizontal beam width of small craft radars will typically be from 2.0 degrees to 8.0 degrees, and the vertical spread as much as 20 to 30 degrees. Rotation of the scanner thus aims a narrow beam around the horizon like a searchlight, from 15 to 40 times per minute depending upon the model.

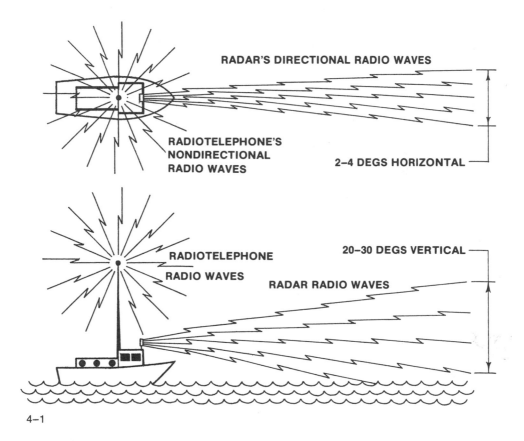

4-1

Transceiver Characteristics

The most complex aspect of a radar is the ability of the transmitter to emit a signal for a fraction of a microsecond, then turn itself off in order for the receiver to detect the echo.

The radio waves, or pulses, are timed in two ways: first, by the length of the burst, measured in millionths of a second and termed *pulse length*; second, by the number of times in one second that the pulses are repeated, termed *pulse repetition frequency*.

For short ranges of up to 3 to 5 miles, short pulse lengths or bursts of power repeated at a higher frequency give better definition. On longer ranges, it is desirable to use longer pulse lengths in order to deliver the greater amount of necessary power to go the 15 to 30 miles to the target and a slower repetition frequency rate to avoid overloading the transmitter.

To look more closely at the pulse length and its relationship to the closest distance from a target that an echo can be seen on the scope, it is necessary to recall that radio waves travel at about 162,000 nautical miles per second. That is 984 feet per one-millionth of a second, or microsecond. Assuming a pulse length of 1.0 microseconds, it means that the radio waves will travel 984 feet in one-millionth of a second. Since they go out to the target and are reflected back, the target cannot be closer than one-half the 984 feet—or 492 feet—to be seen on the scope. If it is closer than that the transmitter will still be emitting a signal, and the receiver will be in its off mode. It is for this reason that shorter pulse lengths are necessary in order to see a target at ranges less than 492 feet. The table below shows the theoretical minimum ranges, but not always realizable in actual practice, for various pulse lengths.

Pulse length in microseconds	Pulse length in feet	Nearest distance in feet
1.0	984	492
0.5	492	246
0.25	246	123
0.1	98	49
0.05	49	25

Some radars have a number of pulse lengths and repetition frequencies, automatically matched for optimum performance to the range selected by the navigator. One model, when on the 1½ mile range has a pulse length of 0.08 microseconds and a pulse repetition frequency of 3000 times per second; when on the 18 mile range the pulse length is extended to 0.5 microseconds and the repetition frequency reduced to 1500 times per second. Some longer range equipment, capable of up to 48 miles range, will have pulse lengths as long as 0.75 to 1.0 microsecond and repetition rates as low as 750 to 850 times per second.

To somewhat oversimplify the foregoing discussion, length and frequency of each transmitted radio pulse can be summarized by saying: distant targets are best seen if the pulse is long (such as 0.75 to 1.0 microsecond) to lend strength to the radio waves when they reach the target, and occurs fewer times per second (such as 850) to avoid overworking the transmitter; whereas for short-range targets, the pulses must be short (such as 0.05 to 0.08 microseconds) in order to avoid overlapping the outgoing signals with the returning echoes, and the repetition rate more frequent (such as 3000 per second) to paint a sufficiently bright echo on the radarscope. Radars with a single pulse length and pulse repetition frequency are seldom offered with ranges of more than 12 to 15 miles, because the values selected must be a compromise between the more desirable multiple pulse lengths and repetition frequencies necessary for optimum performance at both short and long range. One typical make with a maximum range of 12 miles uses a pulse of 0.1 microseconds and a repetition frequency of 1500 per second.

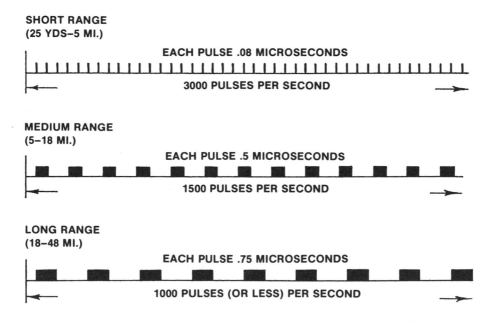

SHORT RANGE
(25 YDS–5 MI.)

EACH PULSE .08 MICROSECONDS

3000 PULSES PER SECOND

MEDIUM RANGE
(5–18 MI.)

EACH PULSE .5 MICROSECONDS

1500 PULSES PER SECOND

LONG RANGE
(18–48 MI.)

EACH PULSE .75 MICROSECONDS

1000 PULSES (OR LESS) PER SECOND

4–2

Receiver Characteristics

The receiver portion of the transceiver merits a few words of discussion. It is the portion of the entire radar system that must accomplish the near impossible. The receiver must be able to paint an echo on the scope from a distant target that has the strength, or lack of it, estimated at one-million-millionth of a watt. Even the strongest echo from a large ship nearby is estimated at only one-ten-thousandth of a watt in power. To compound the problem faced by the receiver is the necessity to amplify the weak echoes and level all signals to a common value for presentation on the scope. A reasonable relationship must be achieved between weak signals returned from a buoy at 4 miles and a strong signal from a ship at 2 miles.

To accomplish these difficult tasks the receiver must be highly sensitive, and the sensitivity is substantially controlled by the "mixer" section. Many developments have been made in recent years in the design of the mixer, one of them being balanced circuitry and the use of two crystals, in place of a straight-through circuit and a single crystal.

Two benefits come from an advanced type receiver. First, the noise factor will be low ("noise" in radar being a counterpart of "hiss" in a radio receiver and manifested as glow on the radarscope that can partially obscure the echoes from targets). Second, a lower mean power output transmitter can be used without any reduction in either brightness or clarity of targets.

Antenna Considerations

As mentioned earlier, the antenna assembly can consist of (1) a radome enclosing the scanner and the transceiver, or (2) an open scanner with the transceiver directly below it, or (3) only the scanner with its driving motor, as illustrated in Fig. 4–3.

The driving mechanism for the scanner is a straightforward motor, gear box, and belt drive to give a rotational speed from 15 to 40 revolutions per minute. Virtually all scanners on modern radars, whether enclosed or open, are of the slotted waveguide type. As the name implies, such an antenna is a horizontal section of waveguide with precision cut slots through which radio waves are emitted and echoes received. Those that are open have a watertight envelope enclosing them, generally made of fiberglass. Some have fiberglass only across the slotted section, and use metal for the rest of the streamlined enclosure. Those that are enclosed are protected by the fiberglass radome which covers the scanner and transceiver.

Width (or span) of the scanner's aperture is the major controlling factor of the radar's range, quality of picture and bearing discrimination. If an open scanner is used, and there is a choice of span widths, a 4-foot antenna is preferable to a 3-foot one; 6-feet is preferable to 4-feet, etc. Beam width varies inversely with span; on an average, the 3-foot antenna will produce a beam width of 2½ degrees, 4-foot a width of 2 degrees, and a 6-foot span a width of 1½ degrees. The narrower the beam the greater will be the ability of the radar to discriminate between two targets that may be close together, whereas a broad beam will cause targets to merge together if they are a mile or more distant.

4–3a. Typical radome antenna system with rotating scanner and transceiver within the fixed dome.

4–3b. An open antenna of greater span than can be used within a radome, mounted above the case that encloses the drive motor and transceiver.

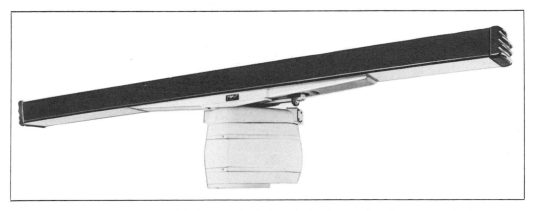

4–3c. The open antenna with drive motor below it of a three-unit radar system which has its transceiver mounted some distance from the antenna with waveguide connecting them.

The vertical beam width of most small craft radars ranges from 20 to 30 degrees. This spread is necessary to compensate for the greater rolling characteristics of small boats as compared to large vessels, where the vertical beam width may be as little as 10 to 15 degrees.

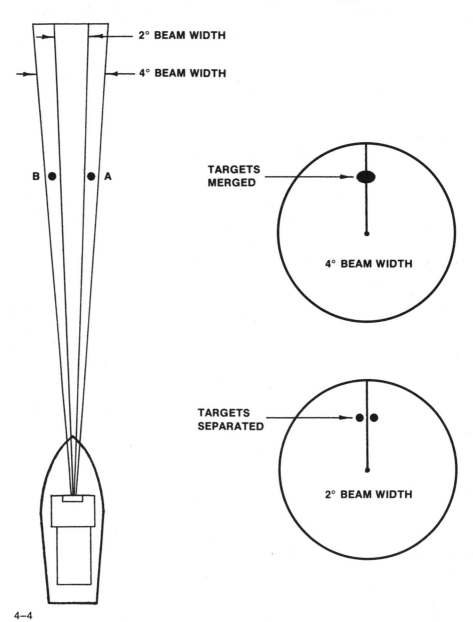

4-4

Display Console

After a short period of "warm-up" the radarscope will show three things: rotation of a faint sweep trace synchronous with rotation of the antenna, a momentary heading flash as the sweep passes dead ahead, and concentric range rings. Around the scope will be the various operating controls.

Assume that the radar is set to a 4-mile range scale, and that there are three range rings spaced one mile apart. The outer perimeter of the scope would be 4 miles. If there were no targets within the 4-mile range, the sweep would be seen rotating at the exact speed of the antenna. The afterglow on the scope created by the sweep would be of a low level. However, there is a bright spot on the faintly illuminated 2-mile range ring, and the azimuth shows it to be 40 degrees to starboard. The persis-

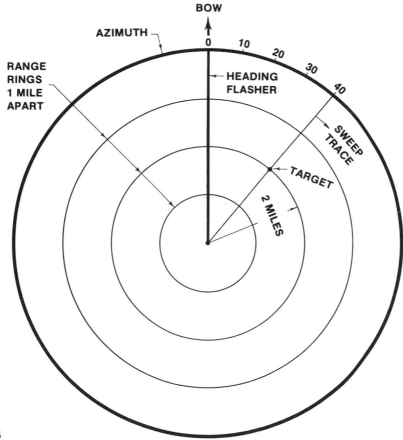

4–5

tence of the scope will keep the echo partially illuminated until the scanner reilluminates it, on its next revolution. On each successive sweep the target will move depending upon its course or speed if it is a boat underway, or upon your speed and course if it is a buoy or boat at anchor.

Changing range scale of the radar to 10 miles, for example, might bring within range the shoreline, and on that scale have only a single range ring at 5 miles. The target that appeared on the second range ring when on the 4 mile range (1/2 the distance out from the center of the scope) would now appear much closer to the center. Proportionally it would be 1/5 the distance out from the center toward the 10-mile perimeter of the scope. See Fig. 4-6.

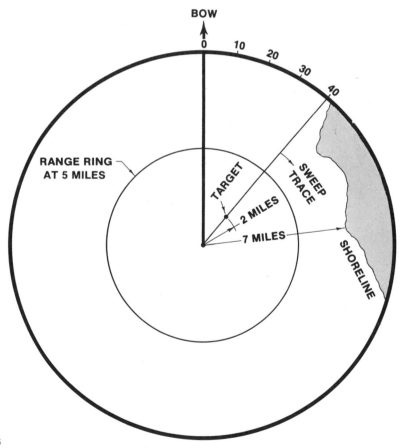

4-6

Simplicity versus Sophistication

Operating controls are a part of the display unit. Assuming the radar has a range of 10 miles or more, there may be as many as a dozen or more controls and switches. The minimum required controls should include anticlutter to reduce sea-return, gain adjustment for optimum sensitivity, tuning adjustment to obtain peak performance, range ring on/off switch, range scale, and brilliance controls. If the radar has a rotatable cursor (versus a fixed engraved grid), there must also be a means of rotating it to determine the bearing of a target.

Some radars may include controls to reduce clutter from rain in addition to sea-return, tuning eyes or performance monitors, range ring and bearing cursor illumination controls, standby switch, and alignment control if it is necessary to align the heading marker to dead ahead when the radar is first turned on.

In considering the number of optional ranges, in miles, it is preferable to have a multiplicity of them for the shorter distances, such as for 1/4, 1/2, 1 and 3 miles. The longer ranges are generally spaced farther apart. Having a number of short ranges can be extremely helpful when entering harbors, or when in heavily congested traffic areas because the entire scope can be used to display nearby targets within the selected range. For example: with a radar having a 7-inch diameter scope (3½-inch radius from center to perimeter) and its shortest range scale is 1 mile, a target at 1/4 mile would be 7/8 inch out (1/4 of 3½ inches) from the center of the scope. If there were other targets between that one and the radar equipped boat, the distance on the scope between them would be something less than 7/8 inch, and they might tend to merge together. On the other hand, with a range scale as short as 1/2 mile on the 7-inch scope, the target at 1/4 mile would be 1¾ inches out from the center. The other targets between it and the boat would be spaced farther apart, for easier tracking and observation.

Another important factor in considering the scope is its diameter. Obviously the larger the diameter, the easier it is to measure distance to or between targets. It is comparable to looking at an enlargement of a photograph versus a contact print from a small negative. Economics enter into the matter and radars with 9-inch-diameter scopes, or larger, cost more than those with smaller ones. Magnifiers can be used to increase the apparent diameter of the presentation from the smaller scopes and are available for some makes. When equipped with a magnifier, however, the navigator must look directly and squarely into the scope to avoid distortion of the picture.

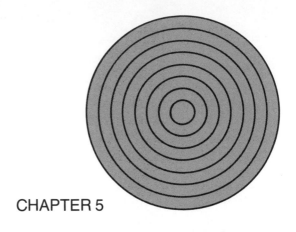

CHAPTER 5

FACTORS AFFECTING RADAR PERFORMANCE

An entire book could be written on this subject, but the majority of it would deal with equations, formulae, and higher mathematics to prove that weather, pulse length, pulse repetition frequency, receiver sensitivity, antenna span, and radiated power are the principal factors affecting performance. Hence, this discussion is confined to the end results of these factors, rather than the theory behind them.

Weather and Atmosphere

As has been stated earlier, the maximum range capability of a radar depends on the height of the antenna and of the target, whether the latter is a buoy, boat, supertanker, island or headland. Related to these heights is the radiated peak output of power. Assuming an adequate amount of power, radar should be able to show an echo on the scope of a target that is approximately 15 percent more distant than the horizon. Beyond that distance the height of the target will control the number of additional miles it can be seen—under standard atmospheric conditions.

There are atmospheric phenomena, however, that can extend the normal range of radar by as much as 30 to 50 percent. Extremely humid air lying as a shallow blanket over the water has the effect of ducting the radio waves and bending them over the horizon. It is termed "super refraction."

Similarly, a layer of warm dry air elevated some distance above the water during a temperature inversion, can also extend the range. The upper portion of the vertical beam will strike the underside of the layer and be reflected downwards to extend the range. Both of these conditions are more apt to occur when the air temperature is higher than the water temperature. Conversely, range can be reduced when the air temperature is lower than the water temperature and neither the high humidity ducting or temperature inversion is present. This is termed "sub-refraction."

Heavy rain squalls, hail, and to a lesser degree snow, can reduce the range. Targets within the squall, and targets beyond, may suffer a reduction in the intensity of their echoes as displayed on the scope. This is due to the rain, hail, or snow absorbing a portion of the radiated power from the radar before the radio waves hit the targets, leaving a lesser amount of power to be reflected from them. Particularly when operating at night and the presence of a rain squall is not visible to the navigator, the wooly appearance of the squall on the scope can give forewarning. A change in course might be indicated to go around the heart of the storm.

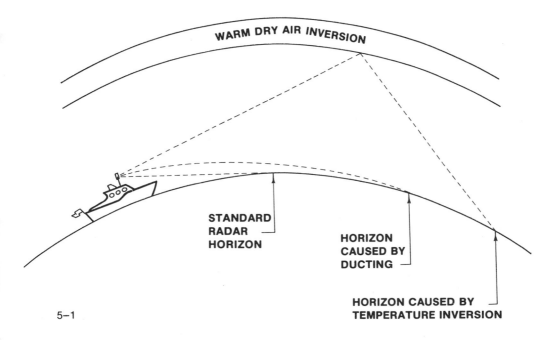

5–1

The sensitivity of some radars is so high that extremely dense banks of fog can be detected on the scope many miles before entering the fog area. Fog does not, however, have the tendency to materially weaken the echoes from targets that lie within or beyond the fog.

Sea-return can also impair the ability of a radar, when on the shorter range scales, to see a target. Cresting waves create echoes that appear as splotches on the scope, out sometimes as far as 1 to 4 miles from the boat. When a small boat or buoy is within that range, they may appear as echoes only intermittently because they become lost in the sea-return. Most makes of radar have a control that will reduce the intensity of the sea-return, which sharpens the echoes from targets. Sea-return is most pronounced in the windward quadrant of the scope. Care should be exercised in using the sea-return control so that it is not reduced to a point that a legitimate target cannot be seen.

Maximum Range

The most important factors are pulse length and repetition frequency. To achieve sharp presentation on the scope of an island 25 miles away requires a strong signal hitting the island in order to be returned. Obviously it requires more power than to get a response from an island only 5 miles away. In the first case the radio waves must travel 50 miles (out and back), and in the second case only 10 miles. In a vacuum there would be no difference in the strength of the returned echoes, but when travelling through atmosphere the radio waves decrease in strength the further they have to travel.

To deliver an optimum amount of power for long range operations (such as 25 to 50 miles) the length of the pulses must be longer than those needed for short range operations. Typically pulse lengths of 0.75 to 1.0 microseconds are used. These are in comparison with shorter pulse lengths of 0.05 to 0.10 microseconds for shorter ranges.

Coupled with the longer pulse lengths to deliver a greater amount of radio energy at the distant target is generally a lesser number of pulse repetitions per second. In radars that have multiple pulse lengths and repetition frequencies, the two are automatically matched for maximum performance when changing from short to longer range scales. On some of the more sophisticated large ship radars there may be as many as four different combinations of pulse lengths and repetition frequencies for extremely short-, short-, medium-, and long-range operations.

Minimum Range

It was noted in Chapter 4 that to see a nearby target it is absolutely necessary to use short pulse lengths. A 0.05 microsecond pulse travels 49 feet in that small period of time. Under ideal conditions, it will display an echo on the scope of a target 1/2 that distance away (25 feet away), and with little difficulty if it is 40 to 50 feet away; a 0.1 microsecond pulse will display an echo when it is between 50 and 100 feet away from the target. A longer pulse of 0.75 microseconds cannot display an echo unless the target is at least 369 feet away, because the echo would be returning before the transmission pulse ended, causing an overlap. Hence the need for short pulse lengths for close-in detection of targets.

Coupled with the short bursts of radiated power is the need to repeat them more frequently in order to get a bright echo on the scope. This dictates the use of faster repetition frequency rates of from 1500 to 3000 or more per second, which has the effect of hitting short range targets with more average power, as compared with 850 to 1000 per second for long-range operation.

Range Accuracy

Most pleasure craft radars have an accuracy in measuring distance (range) of from 1 to 2 percent of the range scale in use. Some manufacturers express accuracy in yards. In either case, the inherent range accuracy of the radar is of less importance than the topography of the headland, island or landmark on which a distance measurement is being made. A second factor is the inability of the observer to precisely interpolate distance between two range rings; a shoreline appearing midway between a 6 and 12 mile range ring might be read by one person as 9.2 miles and by another as 9.8 miles. That difference is substantially more than the inherent range accuracy of the radar.

More accurate range measurements can generally be made by using the shortest range scale that still shows the headland or target. For instance, if on the 18 mile range scale the target's echo appears about 1/4 the distance out from the center of the scope, it might be interpreted as 4½ miles. On the other hand, if a range scale of 6 miles is available, with a range ring at 3 miles, the distance to the target can be more accurately determined because of the enlarged size of the presentation.

Bearing Accuracy

There is greater opportunity for an error in judging an echo's bearing on the scope than its range. A major factor is the antenna span. As was pointed out earlier, an antenna with an aperture width of 30 inches will radiate a horizontal beam approximately 3 degrees in width; a 36-inch scanner will have a beam width of about 2½ degrees; a 48-inch scanner, approximately 2 degrees; as compared to a 12-foot scanner which may have a beam width of only 0.65 degrees.

With a beam width of 3 degrees from a small antenna, when the leading edge of the beam hits a headland, echoes start coming back to the receiver; and during the time that the full width of the beam is on the headland until the trailing edge of the beam is past it, echoes will continue to be returned. The result is that the echo shown on the scope will be of considerable width. When the cursor is laid over the echo from the headland, to get a relative bearing, the question is what part of the echo is the correct bearing. An operating procedure that reduces the width of the echo on the scope is reduction of the "gain" control to a point that the headland is just barely visible on the scope. By rotating the cursor to the middle of the echo a more accurate bearing can be read on the azimuth.

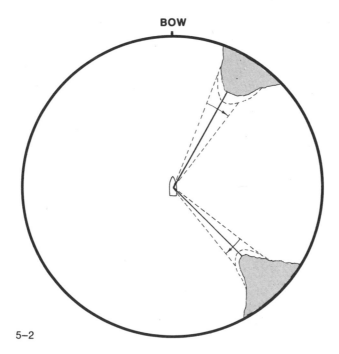

BOW

5–2

Another cause of bearing inaccuracy is the inability of small craft (versus large oceangoing ships) to stay precisely on course. Even with only a moderate sea the boat may be swinging from 5 to 10 degrees during the time that a bearing is being taken. This will, of course, adversely affect the accuracy of the bearing. Under such conditions an average of a number of bearings must be used.

Radar Interference

A final, but not too serious factor affecting the performance of a radar, is interference from other radars. Interference may come from another ship close by, a powerful shore-based radar, or from a ship many miles away whose radar beams are being reflected from an inversion layer. It will be most pronounced if its frequencies and repetition rates are the same as your radar. On the scope such interference appears as radial spokes irregularly spaced, and not reappearing at the same exact positions each time the sweep rotates. The intensity of the spokes, which sometimes may only be dots that tend to form a spoke, are generally less than echoes from true targets such as other ships or nearby buoys. On radars with a number of pulse and repetition rates the interference can frequently be reduced by changing to a different range scale.

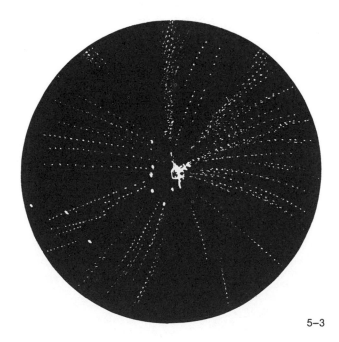

5–3

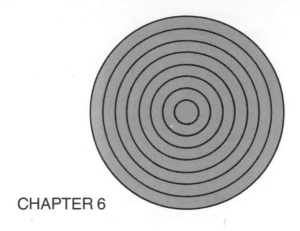

CHAPTER 6

RADAR FOR COASTAL NAVIGATION

Before starting on a coastal cruise during which radar will be the principal navigation tool, consideration is due the types of targets that the radar will be seeing. Some targets are more radar-conspicuous than others; reflective characteristics vary to a considerable extent, affecting the usefulness of radar for navigation and piloting.

Buoys

Unlighted buoys are the most difficult to "see" on radar because they are cylindrical in shape and have no flat surfaces to reflect the radio waves. With a typical radar rated at 3 kw peak output, for example, such buoys may not paint an echo on the scope until you are within 1 mile of them, and then only when the range scale being used is not more than 2 or 3 miles. A higher output radar might pick up the same buoy at 1½ miles.

Lighted buoys with structures to support the light usually have some flat surfaces that will increase their echo-producing capability. They will come into view at from 1½ to 2 miles, generally. If there is a sea running, however, the instability of the buoy coupled with sea-return may cut that distance by one-half.

The more important buoys at harbor entrances, and sometimes along the channel leading into a harbor, generally have a radar reflector as part of its structure. On the chart they will be identified as *RaRef*, meaning radar reflector buoy. Their reflective qualities are from 2 to 4 times greater than other buoys, and with reasonably smooth sea conditions may be seen on the scope as much as 3 to 4 miles distant. However, if there is a heavy swell running, they may be in a trough at the time the radar beam sweeps past the buoy, and thus would not produce an echo.

6–1. Typical radar reflector buoy used at harbor entrances and in channels. The four panels extend from the corner frames to the center of the buoy, directly under the light, to produce surfaces that are 90 degrees to each other on all four exposures.

LIGHTED OR RaRef BUOY

6–2

UNLIGHTED CHANNEL BUOY

Breakwaters

Even though a breakwater may be within clear-weather visibility for as much as 4 to 6 miles, many times it will not appear on the scope until within about half those distances. Most breakwaters are built from granite or concrete blocks placed at random one on top of each other. The surfaces of each block, on close examination, are also highly irregular. Except for the few blocks that might have a vertical plane at right angles to the water, the others will be at various angles. The result is that when radio waves hit them the echoes are scattered in many directions, rather than directly back to the radar antenna, producing too weak an echo for the average small craft radar to detect until within a 2 to 3 mile range.

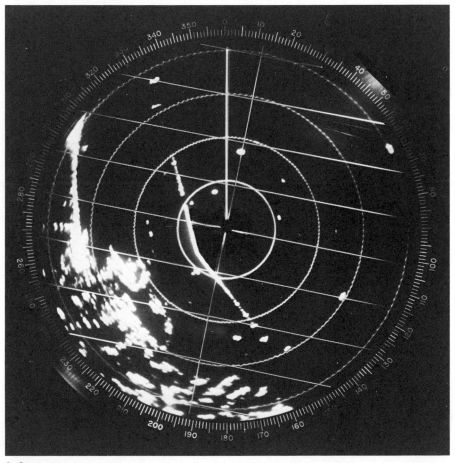

6–3

Fig. 6-3 shows a radarscope, with the radar on its 4-mile range scale. Each range ring is 1 mile apart. On the port side of the boat is a breakwater which fades out of range at about 2 miles even though it is more than 4 miles in length both ahead and astern. On the other hand, land masses at 3 and 4 miles are clearly seen to port. The cursor grid has been rotated 12 degrees to the right of the brighter heading flasher (which represents the bow) and a target at 1¾ miles is directly under the cursor line. Other craft are visible as echoes both inside and seaward of the breakwater. The size of each echo is a clue to the size of the target.

Land Masses and Buildings

Vertical, or nearly so, cliffs will produce stronger echoes than sloping hills; sand dunes between the mainland and the beach are poor reflectors of radio energy. Metal buildings ashore will reflect strong signals, particularly if the aspect of the building is at right angles to the radio waves. Lighthouse structures on points or breakwaters that are hexagonal or square in shape will produce strong echoes when one of the surfaces of the structure is perpendicular to the radio waves, but only weak echoes will return if a corner of the structure is pointed toward the radar.

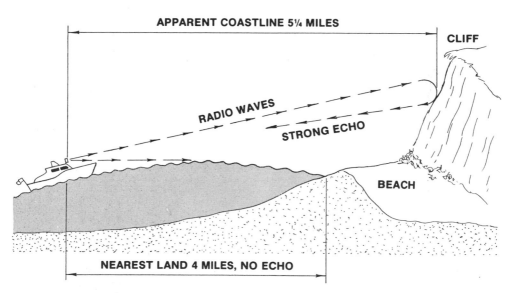

6-4

Echoes from Boats

Steel or aluminum hull boats will produce stronger echoes than wood or fiberglass ones, but the aspect of the boats can be of greater importance than the material of which they are built. Echoes from boats that are bow-on or stern-to the radar will be much weaker than if the boats are broadside. Moreover, the degree to which a boat is rolling will effect the intensity of the echo returned from it. If the radar beam hits its side while the side is perpendicular to the radio waves, the echo will be strong, but if the beam hits the hull while it is rolling, much of the reflected radio energy will be deflected up or down and not paint as strong an echo on the scope.

Metal masts on sailboats are thought by many people to be good reflectors of radio energy, but unfortunately this is not true. As is the situation with cylindrical buoys, radio waves bounce off the mast in all directions, and very little energy is returned toward the radar equipped boat. Nor is metal rigging, regardless of the amount of it, of any assistance in making the sailboat more radar-conspicuous. The only safe route, for either power or sailboats, is to hang a corner-type reflector as high in the rigging as is possible.

A corner reflector, made of sheet metal or metal-foil-covered plywood, can make a bow-on 14-foot dinghy look as large as a 50-foot cruiser when the cruiser is broadside. Echo producing capabilities of reflectors increase greatly with only a small increase in size; one with its inner and outer corners 18 inches apart produces an echo sixteen times greater than one of one-half that size. A further advantage of carrying a reflector is that when at sea, in even moderate swells, the boat may be below the top of the swell and not be seen by a radar equipped boat, but with a reflector in its rigging, regardless of rolling which adversely affects echoes from a hull side, the reflector will paint a strong echo on the scope.

So effective are radar reflectors that the Canadian Ministry of Transport issued a ruling in 1975 that requires reflectors on all nonmetallic hulled pleasure boats commencing January 1978. Between issuance of the order, and that date, commercial boats were required to use reflectors. Even though not *required* on U.S. owned craft, the prudent boatowner should hoist a reflector for his own safety. Aside from making the boat more visible to radar-equipped craft, there is a further advantage pointed out in a recent Coast Guard Local Notice to Mariners. It said "search and rescue aircraft and surface craft use radar to assist in locating disabled vessels. Operators that are the object of a search are requested to hoist a radar reflecting device." One of their officers said "unless we can find a boat in trouble we cannot affect a rescue. Anything that will speed completion of the search shortens the time to make a successful rescue; sometimes this quicker rescue means the saving of lives."

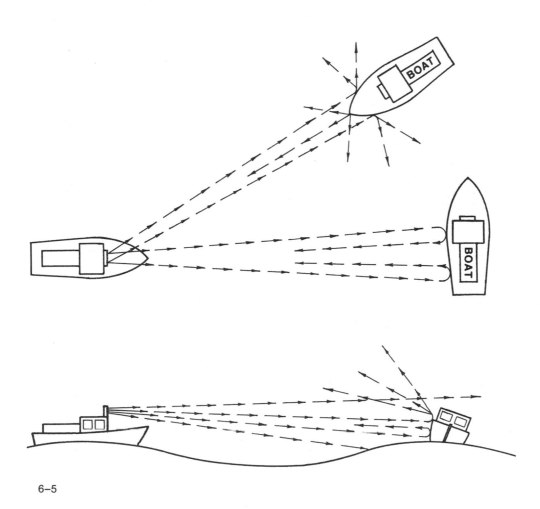

6–5

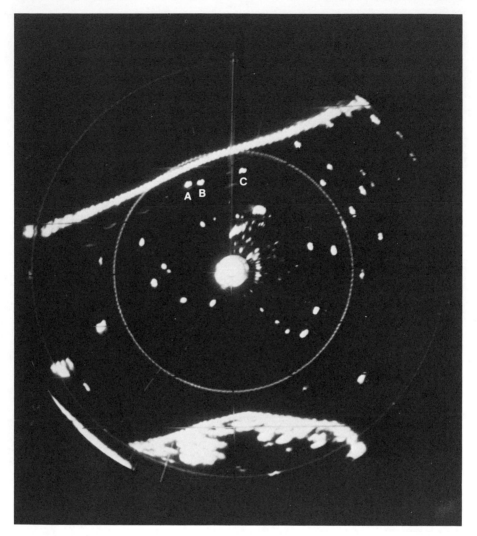

6–6a. Boat A is a 50-foot power cruiser broadside; boat B is a 14-foot fiberglass dinghy with a radar reflector held aloft on an oar; boat C is a 12-foot sailboat with a radar reflector hoisted on a halyard. Without the reflectors the dinghy and sailboat would hardly have been visible on the radarscope.

6–6b. A homemade reflector fabricated from plywood and covered with aluminum foil. When not in use, it can be disassembled and stowed below.

Relative versus Magnetic Bearings

It must be stressed that radar is only a *navigational aid*, and the actual navigating of the boat must be done by the operator. An accurate compass, charts for the area, protractor or parallel rules, dividers or compass-pencil, and a clock are the additional tools needed to do a seamanlike job of coastal navigating with radar.

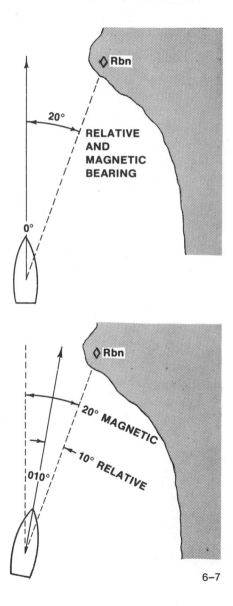

6–7

Before delving into the bearings obtainable by radar, let's refer again to a radio direction finder. Bearings derived with an RDF can be either *relative* or *magnetic*. If the boat is heading due north, or 0 degrees, a bearing on a radiobeacon 20 degrees to the right would produce a relative *and* magnetic bearing of the same number of degrees. But if the boat is on a course of 010 degrees, the relative bearing would be 10 degrees, and its magnetic bearing would be the 010 degrees being steered plus the 10 degrees relative to the course—for a total of 20 degrees magnetic. With radio direction finders having a rotatable azimuth that can be turned to the compass course being steered, the obtaining of a magnetic bearing is accomplished in the single step of getting the null and reading the bearing on the azimuth.

Radars of the type used on most pleasure boats give bearing information only relative to the bow. Assume that the radiobeacon referred to in the preceding paragraph is on a headland, and the boat is on the due north course, the headland would appear on the radarscope as 20 degrees to the right of the heading flasher. The magnetic bearing to it would be 20 degrees. And if the course being steered is 010 degrees, the headland would appear on the scope 10 degrees to the right of the heading flasher. To convert that relative bearing to a magnetic bearing for plotting the single line-of-position requires adding the relative bearing to the course being steered: 10 degrees to the right added to the course of 010 degrees gives the magnetic bearing of 20 degrees.

Bearing information in degrees is obtained from most radars by use of a *rotatable* cursor, which is a transparent disc over the scope with line engraved on it from the center out to the azimuth. A few models have a multiplicity of engraved lines on a *fixed* disc, like the spokes of a wheel, that require interpolation of the degrees of bearing between the lines which may be engraved only each 10, 15 or 20 degrees. The *rotatable* type that allows placing the cursor directly over the echo, is far more accurate, however. Azimuths around the circumference of the scope, on most radars, are fixed and graduated in degrees, clockwise, for their full 360 degrees. When the cursor line is over the echo, its extension to the azimuth allows reading the number of degrees that the target is clockwise from dead-ahead. Up to 179 degrees it would be to the right of the bow; if 180 degrees it would be dead-astern; if 181 degrees it would be to the left. A headland to the left or port side might, for example, be aligned with 322 degrees on the azimuth. To convert that to a relative bearing requires subtracting 322 from 360, to arrive at a bearing of 38 degrees to port.

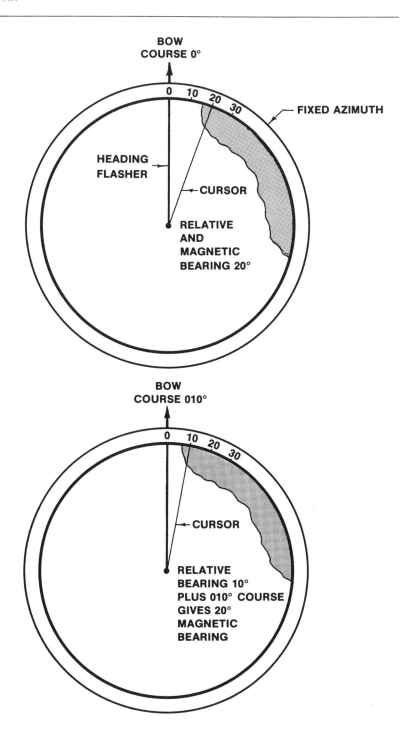

BOW
COURSE 0°

0 10 20 30

FIXED AZIMUTH

HEADING
FLASHER

CURSOR

RELATIVE
AND
MAGNETIC
BEARING 20°

BOW
COURSE 010°

0 10 20 30

CURSOR

RELATIVE
BEARING 10°
PLUS 010° COURSE
GIVES 20°
MAGNETIC
BEARING

6–8

An easier means of determining relative bearing, by providing a direct read-out, is accomplished on some models of radar that combine the azimuth and cursor on a rotatable transparent disc. A small portion of the azimuth is seen through a window at the top of the scope, and is graduated 0 to 180 degrees right and 0 to 180 degrees left. Figure 6–9 shows land to the port side and a headland. The cursor/azimuth disc is rotated to put the cursor directly over the headland. The relative bearing is displayed through the window as being 22 degrees. Since the magnetic course being steered, in this example, is 278 degrees, the magnetic bearing to the headland is 22 degrees subtracted from 278 degrees, or 256 degrees which can be plotted on the chart. Distance to the headland is scaled from the range rings as being 2.6 miles, thus establishing a fix from the single line-of-position.

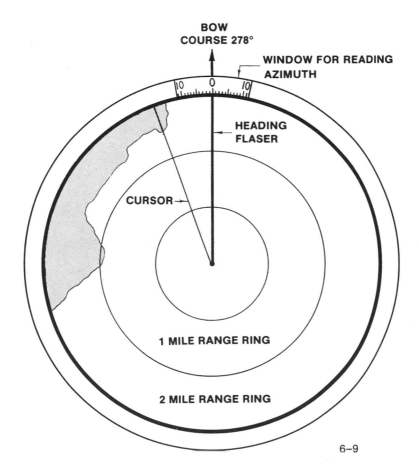

6–9

Fixes by Bearings

The procedure is basically the same as if using a hand-bearing compass, pelorus, or radio direction finder. The chart will show two or three radar-conspicuous targets along the shoreline, or possibly inshore a few miles there may be a prominent hill which is equally useable. Successively relative bearings with respect to the bow will be taken of headlands "A" and "B" and hill "C". Headland "A" bears 30 degrees to the right, headland "B" bears 135 degrees to the right, and hill "C" is 90 degrees to the right. If a course of due north, magnetic, is being steered, the bearings are magnetic and can be plotted on the chart, with the use of parallel rules and the compass rose or by the preferable and more accurate protractor.

Seldom will a three-bearing fix result in all lines-of-position intersecting at exactly the same spot. If they should it is more a coincidence than a reflection of skill by the navigator. Usually a triangle will be formed, within which is the position of the boat. The size of the triangle will be based on the accuracy of each bearing.

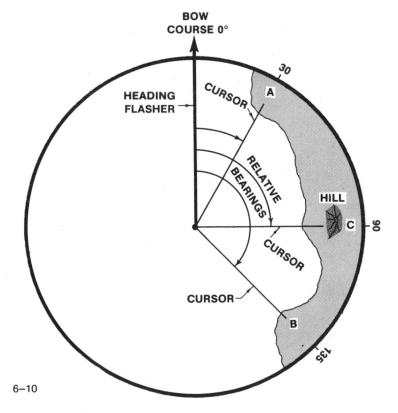

6–10

Should the boat be on a compass course of some other value than due north, or 0 degrees, it is only necessary to add the relative bearing figure to the course to obtain the magnetic bearing if the target is to the right of the bow, or subtract it if it is to the left. For example: assume the course being steered is 325 degrees instead of due north: point "A" would now appear on the scope at a relative bearing of 65 degrees to the right, point "B" would be 170 degrees, and point "C" (the hill) would be 125 degrees. To convert these values to magnetic bearings: add 65 degrees to the course being steered of 325 degrees, giving a total of 390 degrees, from which 360 is subtracted to result in a 30 degree magnetic bearing to point "A"; add 170 degrees to the course of 325 degrees, giving a total of 490, subtract 360 to result in a 135 degree magnetic bearing to point "B"; add 125 degrees to the course of 325 degrees, giving a total of 450, subtract 360 to give a magnetic bearing of 90 degrees to point "C".

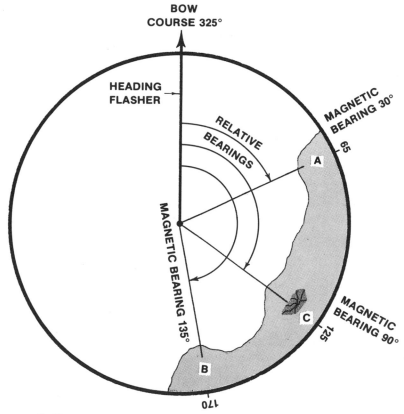

6–11

Fixes by Ranges

Under many circumstances the position of the boat can be more accurately plotted by using the range-determination capabilities of the radar, as contrasted to bearings. This is particularly true if the targets selected are such things as lightships, oil-drilling platforms or steep bluffs. The accuracy of ranges is generally higher than bearings, providing the targets are sharply defined.

Two or three such range measurements can be plotted with a pencil-compass or dividers by noting the distance on the radarscope and setting the dividers to that distance as scaled from the chart. No consideration need be given to the chart course or relative bearings on which range measurements are taken. An arc is drawn for the first range measurement, a second arc drawn for a second measurement, and a third

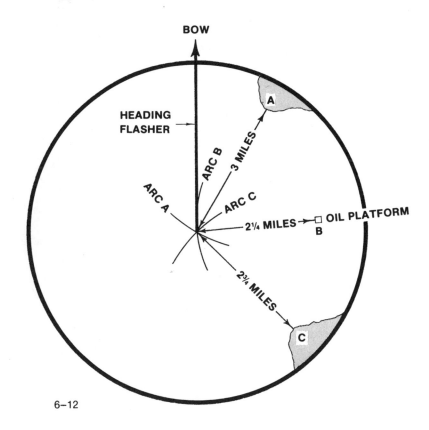

6–12

arc if another target is within radar range. Within the triangle where the three arcs meet is the position of the boat; or if only two arcs are drawn the accuracy of the fix will be within a very close distance to actual position.

Fixes by Bearing and Range

There are some areas of the coast where only one radar-conspicuous target is usable. One of the great capabilities of radar comes into play, under such circumstances. Relative bearing to the target can be read from the azimuth and converted to a magnetic bearing; distance to the target can be read from the range rings. By plotting the single line-of-position with reference to the target, and scaling the distance away from it, gives a very good fix.

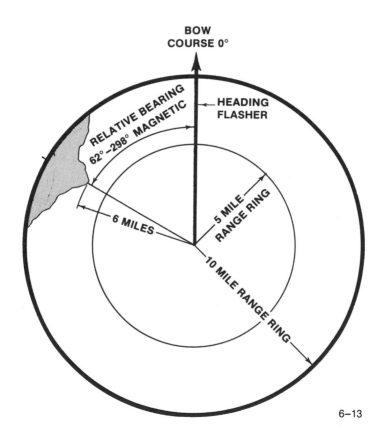

6–13

Fixed or Moving Targets

It is always prudent to form an opinion of whether an echo shown on the scope is a fixed or moving target. Assume that visibility is 2 to 4 miles, or so, and radar is operating; then suddenly the weather deteriorates to the point that radar is needed for both navigation and anti-collision safety. An echo ("A" on Fig. 6–14) shows on the scope 5 miles abeam to the right. At first glance it might be a boat stopped, but it could be going parallel to your course or in the opposite direction. Even worse, it might be on a crossing course. You are still maintaining 10 knots because the visibility is not so poor that you could not stop within one-half the distance of the visibility. At this speed your boat moves 1 mile in 6 minutes.

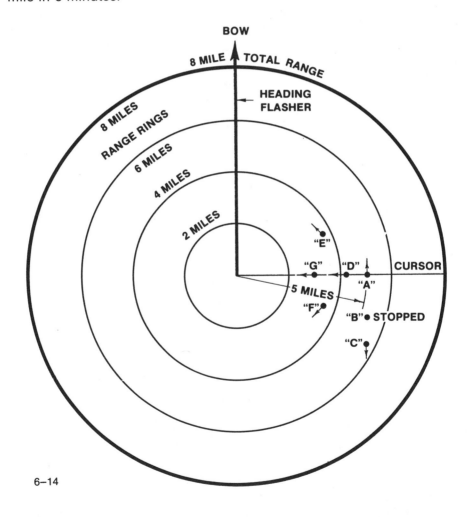

6–14

With only the one target in sight on the scope, the radar's range scale should be switched to the shortest scale that will keep the echo on the scope. This will enlarge the scale and make more discernable any changes of position of the target. The number of range rings that will show on the chosen scale will vary with the make of radar—there could be from one to three or four for use in estimating 1-mile distances which is the distance your boat will travel in 6 minutes.

If the target remains in the same relative position—5 miles to the right and abeam—it means that the *target is moving* in the same direction and at the same speed as your boat. If it falls behind the cursor (to point "B") at a rate of only 1 mile in 6 minutes, the *target is not moving*—it could be a boat at anchor or merely drift-fishing, or it could be a buoy. The relative motion of 1 mile is that of your boat, only. If the distance it moves back (to point "C") is 2 miles in 6 minutes, it means that it is going in the opposite direction at 10 knots. This is derived by the knowledge that you have moved forward 1 mile in 6 minutes, and the other boat has moved an equal distance back in the same 6 minutes.

Should the echo from the target move closer to you (to point "D") it means that it is a boat in motion and on a course that will cross your course. A collision situation could be in the making. A simple way to ascertain the degree of risk is to rotate the cursor to put the etched line directly over the echo when its movement toward you is first observed. In succeeding minutes if the echo ("E") moves forward of the cursor it indicates that the target will cross your bow; if it moves aft of the cursor ("F") it will go astern of you—providing neither craft changes course. If the echo ("G") remains under the cursor line and continues moving closer, a collision will result unless you or the other boat take evasive action.

The same type analysis can be applied to an echo that appears directly, or nearly so, ahead of you. If the distance decreases between you and the echo by 1 mile in 6 minutes (assuming your speed is 10 knots), the relative motion is yours only—the target is not moving. But if you and the target continue to approach each other at a rate of 2 miles in 6 minutes, it means that the target is heading toward you at 10 knots and the relative motion between you and it is at a rate of 20 knots; it also means that its course is approximately the reciprocal of yours.

With your boat and the other approaching each other, nearly bow-on, immediate and bold action should be taken. You have no means of knowing if the other boat is radar equipped, or if it is, that your boat shows on its radarscope. A bold turn to starboard of at least 45 degrees should be made by you, and that course maintained long enough to provide ample sea-room between your boat and the target when you resume the original course. Making a bold course change, versus a small one, adds very little extra distance and if the other boat is radar equipped it will make your course change more noticeable. The extra distance is only that between "B" and "C" as shown on Fig. 6–15.

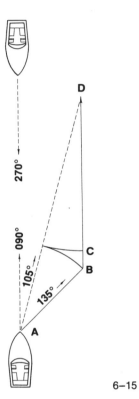

6–15

Speed and Track Made Good

One of the most helpful uses to which radar can be put while cruising along a coast, within radar range of the shore, is to determine actual speed made good over the bottom, and effect of currents with respect to the planned track. Both can be done better and more quickly than by using a radio direction finder because of the greater accuracy in measuring bearing with radar, and its simultaneous presentation of range, or with a hand-bearing compass. The latter, however, requires good visibility to landmarks and at night is usable only if identifiable lights are ashore.

To illustrate a typical case: you are cruising at 10 knots with the intended track laid approximately parallel in a northwesterly direction past headlands A, B, C, and D. Eastward of headland "C" is a bay, and off from headland "D" are rocks extending seaward some distance. The coastline is either shrouded in haze or fog and cannot be seen, or it may be at night during good visibility but there are no lights on the headlands with which to take visual bearings; the waters are deep and beyond the range of the depth sounder.

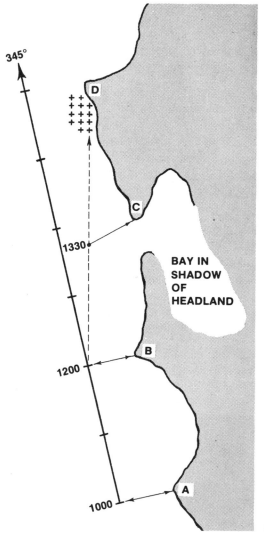

6–16

When abeam point "A" the range of 7 miles is noted and that distance transferred to the chart with a notation of time abeam. While approaching point "B" it is clearly in view on the scope. The distance offshore is observed as 10 miles when abeam, plotted on the chart with a notation of time, and the distance from abeam "A" and "B" is scaled and found to be 20 miles. Since 2 hours had elapsed between the 2 points, the speed made good is the estimated 10 knots—without any help or hurt by currents, winds, or seas; nor has there been a set in or offshore.

When near point "C" both range and bearing is read from the scope and plotted. It places the boat considerably inshore from the intended track, and the distance travelled from point "B" is only 13 miles, during the 1½-hour period. This shows that even though steering the planned course the boat has been set toward shore and the current from the northwest flowing into the bay had slowed the boat over the bottom. Without the benefit of radar, maintenance of the course would ultimately set the boat onto point "D," instead of seaward a safe distance from the rocks.

An immediate change of course seaward to the originally planned track is obviously necessary, and a recalculation of estimated time to arrive at point "D." Whether or not the current from the northwest continues to slow the boat can be determined by the distance covered between time abeam of points "C" and "D."

Anchoring with Radar

A high degree of confidence in radar is a requisite for its use when entering a small cove that is shrouded in fog or during a pitch black night. During daylight and clear weather conditions, the human eye can accurately pick out the middle of the channel, or estimate distances away from buoys and other anchored boats, or relative bearings to a pier or other landmark when picking the spot to anchor. But at night, even without fog, distances and bearings are not easily ascertained by the best of eyes.

Over the years we have developed a few rules for entering small coves during radar conditions. First, we plot courses on the chart and mark the magnetic courses to be steered to approach the cove and the courses into the cove to the point where the anchor will be dropped. We also mark ¼-mile intervals on the course lines from the anchoring point seaward. This helps when correlating the boat's position to what is seen on the scope. Once nearing the final approach to the cove, either my wife or I handle the boat and the other is radar observer; we never change from one position to the other during the final approach. There is frequent exchange of information on what course is being steered, and whether radar shows that a few degrees left or right is needed to offset windage or current. A detail chart of the cove is readily accessible to the radar observer, in order to compare the radar picture with the chart. We have also used this system when intending to use a mooring buoy, rather than anchoring. It does, however, require a minimum of 75 feet visibility through the fog, because our radar cannot see a target when closer than that.

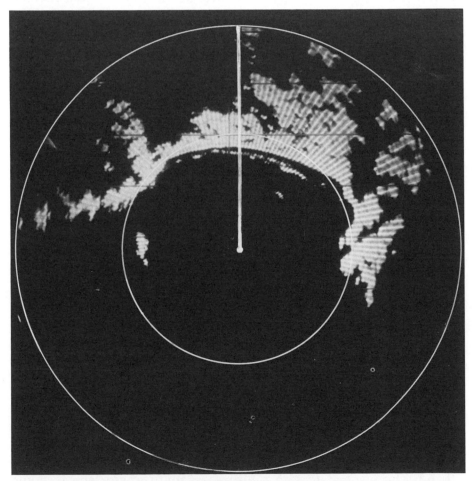

6–17. Approaches to an anchorage with the radar on the 1-mile range. The mid-circle is ½ mile. Note breakers near the beach, the islet just under ½ mile on port beam and rock about 1/3 mile off the starboard bow.

An incident occurred a few years ago involving a fog-bound cove and a nonradar, nonradiotelephone-equipped boat trying to find it. We had anchored in the cove earlier in the day, and an airline captain friend with a new boat was due to meet us during the early evening. Before his estimated arrival time a thick fog moved in and it was impossible to see more than a few yards. Shortly before our friend was due we turned on the radar and watched it for signs of his arrival. Not long after we observed an echo slowly moving across and at right angles to the harbor's axis; then it reversed course and went the opposite direction. This maneuver was repeated a number of times, about ½ mile off the opening to the cove.

It was our assumption that our friend knew his approximate position but was "holding" offshore in the hope that the fog would lift, rather than chancing an approach into the cove in poor visibility. Our son volunteered to take the tender and intercept him, with directions from us to him via walkie-talkie CB radio. The tender was equipped with a compass, and a radar reflector was lashed to an upturned oar. Into the dark night he proceeded seaward with instructions of courses to steer from us by radio, based on his echo on the radarscope. It was no problem to direct him to an intercepting position with our friend's boat. From that point, our tender was tied alongside and with the walkie-talkies for communications we directed him into an anchoring spot near us. This "rescue by radar" is only one of many that we have performed since then, under similar conditions.

Approaching Harbors

When approaching a major harbor that has commercial traffic going into it or out, there are a few more additional precautions that should be taken than when entering a secluded cove. Such harbors as New York, San Francisco, Los Angeles, and areas in the Chesapeake Bay now have traffic lanes leading to or away from them, for inbound and outbound traffic. These are shown on charts and indicate a separation lane between in/out traffic. In the case of the approaches to Los Angeles, the lanes commence as far northwest as Santa Barbara—roughly 80 miles. There are no regulations making it mandatory for a vessel to be within the correct lane, but the prudent navigator of a tanker, freighter, passenger liner, or the smallest pleasure craft will observe them for his own safety. And if it is necessary to cross a lane, it should be made at or near right angles in order to cross it in the least possible time.

At the harbors just named, and many others, there are sea-buoys, generally designated as #1, as differentiated from others marking the channels into the harbors. The latter are progressively numbered 3, 5, 7, 9, etc. if they mark the left side of the channel going into it; or 2, 4, 6, 8 etc. if they mark the right side of the channel. Particularly during poor visibility conditions, the sea-buoys should be approached to put them on the port side when entering the harbor; from the sea-buoy inbound keep on the right side of the channel, as close to the even numbered or red buoys as possible.

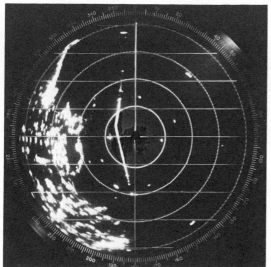 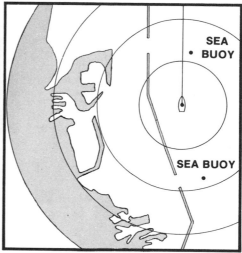

6–18a. Photograph of radarscope showing, to port, portions of the Los Angeles/Long Beach breakwater, the inner harbor, and the land masses further to port. Radar was on 4-mile range scale, with each ring 1 mile apart. The sea buoys seaward of each of the breakwater openings are indicated on the sketch and show as echoes on the scope. Other echoes are boats anchored within the harbors or seaward of the breakwater.

6–18b. Sketch of area in 6–18a taken from a chart.

Never should entry into a harbor be made on a course that crosses the normal steamer lane. Laying out courses and distances as suggested when entering a small cove is just as important when entering a major harbor. Familiarization with the approaches to a harbor, and the detail chart, to form a mental picture of breakwaters, buoys, piers, and channel markers is also extremely helpful. If there are bridges spanning a channel, they will appear as a solid line on the radarscope until you are nearly under them. This is because the radio waves emitted at a vertical angle of up to 30 degrees above the water do not differentiate between an object at water level or one that is elevated. Unless the navigator anticipates the echo from a bridge, it can be very disconcerting because it will appear like a solid breakwater or pier.

A bridge incident occurred to us a few years ago that remains vividly in our memories. We were bound for San Francisco in dense fog; we located the lightship, which was their sea-buoy but now a monster-unmanned-buoy, and took a departure from it to the Golden Gate. The channel buoys from the lightship to the Golden Gate showed clearly on the scope as we came within range of them. When about 1½ miles from the Golden Gate, the shoreline across our course was a solid line on the scope, although the headlands seaward of the Golden Gate were clearly defined. Equally disturbing was the fact that the course being steered to maintain our proper position in the channel was nearly 10 degrees different than plotted, due to the strong cross currents. Not until the foghorn in the center of the Golden Gate Bridge was dead ahead, and possibly 1/4 mile away, did the bridge on the radarscope finally open. Until then the radio waves had been hitting the bridge and producing an echo from its entire span. Only when the upward angle of the radio waves passed under the span did the bridge appear to open.

Practice during Good Weather

There is no better way to develop confidence in radar, and to appreciate the significance of echoes produced by various types of targets, than to practice during clear weather. First practice should be done while anchored in an area where other craft will be passing within a mile or more, and where there are also buoys, headlands, or other fixed targets. Observing both the moving and fixed targets on the scope, then looking at them with your eyes, greatly helps to equate the presentation on the scope to what is actually taking place.

During this exercise, movement of echoes from moving targets can be measured with the use of range rings. Relating the time for an echo to move a certain distance across the scope will indicate the speed that the target is moving. For example, if the target moves 1/2 mile in 3 minutes, it means that it is travelling at 10 knots; if it moves 1 mile in 4 minutes, its speed is 15 knots. Tables to convert time and distance into speed are available in books on navigation.

After developing a degree of proficiency in recognizing fixed and moving targets while the boat is anchored, the next stage of practicing is to cruise along a coastline or through a large harbor in clear weather. With a qualified person on the wheel, and yourself on radar, the effect of your own speed on the fixed targets can be studied; as well as on

6–19. Comparing what is shown on the radarscope with what can be seen from the boat, during clear weather, is an important part of radar training.

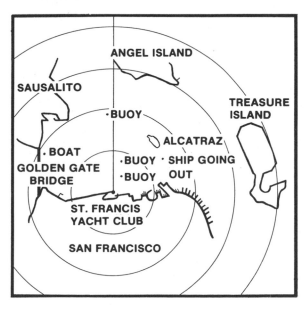

SAUSALITO

ANGEL ISLAND

TREASURE ISLAND

·BUOY

·ALCATRAZ

·BOAT

·BUOY · SHIP GOING OUT

GOLDEN GATE BRIDGE

·BUOY

ST. FRANCIS YACHT CLUB

SAN FRANCISCO

targets that are moving. The practice should include giving the helmsman course changes, based on what the radarscope shows, to avoid
targets; navigating through buoyed channels to develop more proficiency in relating the buoys and other landmarks shown on the chart
to what the radarscope shows.

During these practice sessions, set the radar range to one of the
shorter ranges, such as 2 miles. Note an echo from a boat that may be
only 1/4 mile distant; then change the range scale to a longer one, such
as 10 miles. Two important situations will be disclosed: first, the target
at 1/4 mile will nearly disappear into the center spot and could easily go
unnoticed if the radar was kept on the longer range; second, a number
of echoes at ranges of 5 to 8 miles might be shown. Watching them for
a few minutes shows they are heading toward you. If you had not
switched to the longer range, existence of the other ships approaching
you would not have been disclosed until they came within the 2-mile
range. You would then have little time to assess the risk of collision with
them, and to take evasive action. The moral of this practice is to change
range scales frequently from short to long, or vice versa, to be certain
that all targets within radar range are given an opportunity to be seen.

The third stage in this self-instruction program is to plan a cruise of
some distance to another harbor. Before starting, lay out courses to be
steered on the chart, make notations of distances offshore when passing headlands, etc. While some boatowners tend to avoid marking up
their charts with such notations, it is a false economy and far better to
have all of the navigational information that will be helpful clearly
marked before embarking on the cruise. This will include courses, distances scaled in 1- or 5-mile intervals, relative bearings, and distances
to headlands or buoys when courses are to be changed.

Proficiency in measuring distances to an echo with the aid of the
range rings, and of determining relative bearing of it comes only from
practice. The navigator of a radar-equipped boat should use his radar
as much as possible, regardless of unlimited visibility, to develop capabilities of quickly and accurately determining distance and bearing, so
he will have confidence in the radar and his own abilities when conditions turn foul.

The practice should not be limited to one person of a husband/wife
cruising team; both should be equally able to navigate by radar. Some
persons develop a hypnosis after watching radar for a prolonged time,
and being able to swap duties as radar observer with "conning" of the
boat is highly desirable.

operator and trained in the methods of plotting.

Few small craft have the space near the radar display unit and steering station, or the trained personnel, to do elaborate plotting. Nor is it as important to do so as it is aboard a large ship that is slow to maneuver and may take as much as 3 to 4 miles to bring to a stop. However, an understanding of basic plotting methods, at least to the extent of determining closest point of approach, can be a worthwhile goal.

Rapid Radar Plotting

If the scope of the radar is large enough in size, which it seldom is on the average small craft, the navigator may use the *rapid plotting* method. This is not as accurate or complete as *relative* or *true motion* plotting, but it is adequate in many instances. It is done directly on the safety glass covering the face of the scope, or on a reflection plotter mounted directly over the scope. A grease pencil is used to mark the position of an echo with a notation of time. After three or more marks have been plotted, a line can be faired through them and extended across the heading marker to indicate the closest point of approach. The distance can be interpolated from the range rings.

At night when the hood over the radarscope is not needed to shield light from the scope, there are no problems in marking the positions of echoes. For daytime operations, some large radars have hand-holes on the side of the hoods to facilitate marking the echoes and time. On one of our boats we had a heavy black-out curtain that hung from the ceiling of the pilothouse so that during prolonged daylight radar operations, the observer could comfortably sit in his "radar booth" and could mark on the scope's protective glass.

Relative Motion Radar Plotting

The more accurate method of determining closest point of approach of another vessel is to transfer the information from the radarscope to plotting sheets. Available in pad form from chart houses, they are printed with a 360-degree azimuth corresponding to a compass rose or the azimuth around the scope, and concentric circles which may correspond to the range rings of the radar. That they exactly correspond or do not is of no significance. In the left margin of the sheet are distance scales for 8, 12, 16, and 20 mile ranges; at the right are speed scales; and at the bottom is a logarithmic nomogram for use in computation of speed, distance, and time problems. There is also a box near the top of the speed scales for the navigator to make notes of time, bearing and range of echoes to be plotted. Complete instructions on plotting are described and illustrated on the cover sheet of the pads.

A typical example of the use of the plotting sheet to determine only the closest point of approach is shown in Figure 7-1. At 0900 hours a target appears 7 miles away and at a relative bearing of 45 degrees off your starboard bow. Arbitrarily you assign a distance of 2 miles between each of the four concentric circles printed on the plotting sheet.

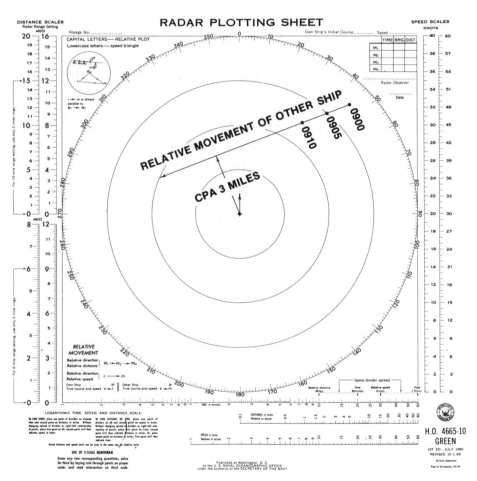

7–1

With a ruler laid from the center of the sheet to 45 degrees on the azimuth, place a mark at 7 miles and a notation of time. Five minutes later you observe that the target is 6 miles away and bearing 40 degrees. You mark that on the sheet; and 5 minutes later it is 5 miles and 34 degrees. A line is drawn through the three plots and extended across the plotting sheet to show the relative motion of you and the other boat. Providing you and the other craft maintain the same speeds and courses, it shows that it will cross your bow.

The closest point of approach can be measured by drawing a line at right angles to the relative motion line of the other boat and to the center of the sheet. In this case, it shows that the distance will be about 3 miles. You should continue to plot its echoes until the boat has crossed your bow and is well to port of you. If it changes course or speed in a way that the CPA would be less, the new plots will immediately alert you to the need for evasive action.

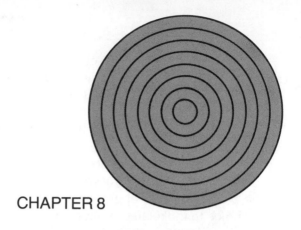

CHAPTER 8

RADAR AND RULES OF THE ROAD

In 1889 representatives of the world's most active seafaring nations met in Washington, D.C., to formulate rules and regulations that all navigators could use for the safety of lives and property at sea. These International Rules of the Road remained unchanged until 1954 when revisions and simplifications agreed upon by all nations were made effective; they were again revised in 1960. The most significant changes made then were relative to conduct of vessels in restricted visibility: a new rule was added, but more importantly, an Annex entitled "Recommendations on the Use of Radar Information as an Aid to Avoiding Collisions at Sea" was included in the rules.

Notwithstanding the capabilities of radar, the use of it gives no new rights to the navigator that were not in the original International Rules of the Road promulgated over three-quarters of a century ago—or in the revisions since that time. As has been stated repeatedly, radar is only an *aid to navigation*. The rules applicable to collision avoidance are still based solely on the abilities of navigators to *see* and *hear* whistle signals of another vessel.

During clear visibility, day or night, the navigator of one boat can quite accurately assess the course and speed of another. Its relative position and the angle at which it is approaching determines which boat is "privileged" or "burdened." The privileged power vessel is required by international law to maintain course and speed; the burdened boat must change speed or course or stop to avoid a collision. When in restricted visibility, your hearing the fog signal of a ship only denotes its presence within audible range, and whether you are privileged or burdened cannot be ascertained until you are within sight of the ship. Therefore, to protect everyone, the regulations acknowledge the aid radar gives to navigators, yet clearly prohibit it from playing a legal role.

Part C of the International Rules of the Road is titled "Sound Signals and Conduct in Restricted Visibility." Excerpted from it is the following:

Preliminary

1. The possession of information obtained from radar does not relieve any vessel of the obligations of conforming strictly with the Rules and, in particular, the obligations contained in Rules 15 (signals) and 16 (speed).
2. The Annex to the Rules contains recommendations intended to assist in the use of radar as an aid to avoiding collisions in restricted visibility.

Speed in Fog

16a. Every vessel, or seaplane when taxiing on the water, shall, in fog, mist, falling snow, heavy rainstorms, or any other condition similarily restricting visibility, go at a moderate speed, having careful regard to the existing circumstances and conditions.
16b. A power-driven vessel hearing, apparently forward of her beam, the fog signal of a vessel the position of which is not ascertained, shall, so far as the circumstances of the case admit, stop her engines, and then navigate with caution until danger of collision is over.

Similarly, if the other boat hears your fog signal forward of its beam, it is also required to stop its engines and proceed with caution, both boats still having some headway even after stopping engines. This Rule applies even if both boats have radar, because radar does not *see* in the same manner that the Rules are written.

When both boats are within visual sight of each other, a meeting, crossing, or overtaking situation can be seen to exist. Which boat is then privileged or burdened can be seen, and the appropriate whistle signals can then be exchanged.

The question of how far the navigators of each boat can *see* and what is *moderate speed* are the crucial points. At this time there are no specifications or definitions of either visibility or moderate speed in the Rules of the Road. Courts have found, however, that the navigator of a boat in fog must be able to stop within one-half the distance of visibility. The boat could be going 30 knots with only 500 feet of visibility, but if it can be stopped within 250 feet the navigator would be within the law, and the 30 knots could be considered a moderate speed. On the other hand, an oceangoing freighter travelling at only 5 knots could not possibly be stopped within 250 feet, and hence its 5 knot speed would not be a moderate speed. Further complicating the problem is that it is often very difficult to gauge the exact range of visibility, and the fact that fog varies in intensity very rapidly—1 minute visibility might be 1/2 mile and the next minute less than 100 feet.

A relatively simple way to get an approximation of visibility, during daylight, is to toss a paper box overboard. If you are cruising at 10 knots, in 1 minute you will travel about 1000 feet. Should the box disappear into the fog in 1/2 minute, the visibility is about 500 feet; or in 2 minutes, the visibility would be about 2000 feet.

Another Rule worthy of citing is Rule 18a of Part D "Steering and Sailing Rules."

Power Driven Vessels Meeting End On
18a. When two power-driven vessels are meeting end on, or nearly end on, so as to involve risk of collision, each shall alter course to starboard, so that each may pass on the port side of the other. This Rule only applies to cases where vessels are meeting end on, or nearly end on, in such a manner as to involve risk of collision, and does not apply to two vessels which must, if both keep on their respective course, pass clear of each other.

As with the other Rules, this one applies to navigators able to *see* each vessel. Nevertheless, the principle is equally applicable to those using radar and is specifically referred to in the Annex to the Rules proposed at the International Conference on Safety of Life at Sea held in 1960. Quoted below is the Annex:

Recommendations On The Use Of Radar Information
As an Aid to Avoiding Collisions at Sea
1. Assumptions made on scanty information may be dangerous and should be avoided.
2. A vessel navigating with the aid of radar in restricted visibility must, in compliance with Rule 16a, go at a moderate speed. Information obtained from the use of radar is one of the circumstances to be taken into

account when determining moderate speed. In this regard it must be recognized that small vessels, small icebergs, and similar floating objects may not be detected by radar. Radar indications of one or more vessels in the vicinity may mean that "moderate speed" should be slower than a mariner without radar might consider moderate in the circumstances.

3. When navigating in restricted visibility the radar range and bearing alone do not constitute ascertainment of the position of the other vessel under Rule 16b sufficiently to relieve a vessel of the duty to stop her engines and navigate with caution when a fog signal is heard forward of the beam.

4 When action has been taken under Rule 16c to avoid a close quarters situation, it is essential to make sure that such action is having the desired effect. Alterations of course or speed or both are matters as to which the mariner must be guided by the circumstances of the case.

5. Alterations of course alone may be the most effective action to avoid close quarters provided that:

a—there is sufficient sea room.

b—it is made in good time.

c—it is substantial. A succession of small alterations of course should be avoided.

d—it does not result in a close quarters situation with other vessels.

6. The direction of an alteration of course is a matter in which the mariner must be guided by the circumstances of the case. An alteration to starboard, particularly when vessels are approaching apparently on opposite or nearly opposite courses, it generally preferable to an alteration to port.

7. An alteration of speed, either alone or in conjunction with an alteration of course, should be substantial. A number of small alterations of speed should be avoided.

8. If a close quarters situation is imminent, the most prudent action may be to take all way off the vessel.

In summarizing the legal status and application of radar, it must be repeated that its use *does not* give the navigator a single additional privilege that is not now provided for in the International (and Inland) Rules of the Road which are based exclusively on visual sight and the use of whistle signals. And, a vessel being visible on a radarscope does not constitute *sight* in the eyes of maritime law.

An additional factor warrants mentioning: courts have held that if a boat is equipped with radar it is the obligation of the navigator to use it during periods of restricted visibility. Should a grounding or collision occur, due to fog, darkness, or at other times of bad visibility, and the radar had not been in use as an aid to navigation, the operator can be charged with negligence. He may be held partially or totally liable for damages resulting from the accident.

NOTE: During 1972 the Inter-Governmental Maritime Consultative Organization, under the auspices of the United Nations, developed new International Regulations for Preventing Collisions at Sea. They were finally ratified by the United States and put into effect July 15, 1977. The U.S. Coast Guard's publication *Navigation Rules* (CG-169) details the Regulations with cross-references to the 1960 International Rules to show which ones are new, revised, reworded, or identical to the former ones. Among the changes are new rule numbers, and use of the terms "stand-on vessel" in place of "privileged vessel," "give-way vessel" in place of "burdened vessel," and "safe speed" in place of "moderate speed." In many respects the 1960 Rules and Annex quoted on preceeding pages are more explicit than the new rules and for that reason their wording has been retained in this book.

Insofar as the use of radar for collision-prevention is concerned, the new rules are essentially unchanged from the 1960 rules although they have been renumbered, rearranged, and in some cases reworded to fit the format of the new rules.

Quoted below are excerpts from Part B—Steering and Sailing Rules that became effective July 15, 1977:

Section I—Conduct of Vessels in any Condition of Visibility
Rule 6   Safe Speed.
Every vessel shall at all times proceed at a safe speed so that she can take proper and effective action to avoid collision and be stopped within a distance appropriate to the prevailing circumstances and conditions.
In determining a safe speed the following factors shall be among those taken into account (by vessels with operational radar):
(i) the characteristics, efficiency, and limitations of the radar equipment;
(ii) any constraints imposed by the radar range scale in use;
(iii) the effect on radar detection of the sea state, weather, and other sources of interference;
(iv) the possibility that small vessels, ice, and other floating objects may not be detected by radar at an adequate range;
(v) the number, location, and movement of vessels detected by radar;

(vi) the more exact assessment of the visibility that may be possible when radar is used to determine the range of vessels or other objects in the vicinity.

Rule 7   Risk of Collision

(a) Every vessel shall use all available means appropriate to the prevailing circumstances and conditions to determine if risk of collision exists. If there is any doubt such risk shall be deemed to exist.

(b) Proper use shall be made of radar equipment if fitted and operational, including long-range scanning to obtain early warning of risk of collision and radar plotting or equivalent systematic observation of detected objects.

(c) Assumptions shall not be made on the basis of scanty information, especially scanty radar information.

Section III—Conduct of Vessels in Restricted Visibility

Rule 19

(a) This Rule applies to vessels not in sight of one another when navigating in or near an area of restricted visibility.

(b) Every vessel shall proceed at a safe speed adapted to the prevailing circumstances and conditions of restricted visibility. A power-driven vessel shall have her engines ready for immediate maneuver.

(c) Every vessel shall have due regard to the prevailing circumstances and conditions of restricted visibility when complying with the Rules of Section I of this Part.

(d) A vessel which detects by radar alone the presence of another vessel shall determine if a close-quarters situation is developing and/or risk of collision exists. If so, she shall take avoiding action in ample time, provided that when such action consists of an alteration of course, so far as possible the following shall be avoided: (i) an alteration of course to port for a vessel forward of the beam, other than for a vessel being overtaken; (ii) an alteration of course towards a vessel abeam or abaft the beam.

(e) Except where it has been determined that a risk of collision does not exist, every vessel which hears apparently forward of her beam the fog signal of another vessel, or which cannot avoid a close-quarters situation with another vessel forward of her beam, shall reduce her speed to the minimum at which she can be kept on her course. She shall if necessary take all her way off and in any event navigate with extreme caution until danger of collision is over.

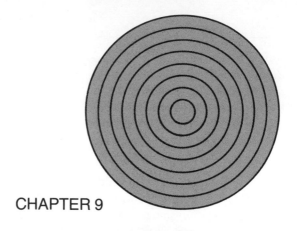

CHAPTER 9

INSTALLATION CONSIDERATIONS

The greatest boon toward easing installation of radars on small craft has been the marked reduction in weight of the principal components. Whereas only a few years ago the antenna systems alone weighed in the neighborhood of 150 pounds, today few weigh more than 100 pounds, and many weigh substantially less. Similarly the weight and size of display consoles has been greatly reduced, due largely to transistorization and printed circuit boards.

Antennas on Sailboats

There are a few more problems and factors that must be considered, when installing radar on small boats. If on a yawl or ketch the antennas are generally mounted on the mizzen mast, high enough to clear the mainsail and backstays. Most owners will select a radar having an enclosed scanner to preclude any risk of lines or sails fouling the scanner. With the transceiver also within the radome, a multistrand cable can be used (versus waveguide) to connect the antenna system with the display console.

Since it is not uncommon that masts must be removed from time to time for repairs, it is vitally important that near where the mast passes through the deck a watertight junction box be installed for the umbilical cable. In

9–1a. Mizzen mast mounted radome of Decca Model 050 is about 7 feet above the deck and 10 feet above the waterline.

9–1b. The display console for Decca Model 050 is on a hinged platform. When in use, directly forward of the chart table, it is swung out from its stowed position over the upper bunk.

this manner, when it is necessary to remove the mast, the connections can be undone from the terminal strip within the junction box. Use of a watertight quick disconnect coupling is an alternative to use of a junction box. It would otherwise be necessary to disconnect the cables at the display unit, and pull them through compartments and a watertight packing gland at the deck adjacent to the mast.

Many owners of racing sailboats strenuously object to any extra weight above the decks, particularly ten or fifteen feet up a mast. Obviously elevation increases the range of a radar but, as noted earlier, doubling the height from 10 to 20 feet increases the radar horizon by a distance of only 1½ miles.

On sloops it is not uncommon to mount the radome enclosed antenna on a spreader. This requires a spreader of more than usual dimensions and rigidity to insure a steady mounting base for the antenna. Some sloop owners have installed masts 6 to 10 feet high near the transom, on top of which is the antenna. While this installation looks "odd," it is nevertheless entirely practical.

An interesting installation was made a few years ago aboard a 43-foot yawl. The owner wanted an open scanner antenna, but at a location that was out of the way, as low as possible in relation to the deck, and in a position that would not seriously compromise radar performance nor interfere with the running rigging. The base of the antenna assembly was modified slightly and bolted directly to the main cabintop about 5 feet aft of the mast. Height of the scanner measured about 5½ feet above the waterline.

9–2. Tubular tripod on the transom of this sloop supports the radome-enclosed transceiver/antenna.

The eight-foot dinghy normally carried on the cabin top was modified only to the extent that the seats were removable. This permitted inverting the dinghy over the antenna assembly. The same chocks on the cabin top formerly used to hold the dinghy in position were retained. Since the

9–3. Antenna/transceiver assembly is mounted on cabin top of this sailboat, just aft of skylight. Fiberglass dinghy is stowed over it when the dinghy is aboard. Span of the antenna is enough less than the beam of the dinghy to permit its rotation.

9–4. Fiberglass dinghy totally covers the antenna system.

dinghy was of fiberglass construction, it had no adverse affect on the radar's performance, and for all practical purposes was the counterpart of a radome that would be used on a smaller radar.

A different, but similar, method of locating the antenna near deck level is shown in Fig. 9–5. The radome is bolted directly to the main cabin skylight, yet permits opening of it for ventilation and does not adversely restrict natural light through the frosted plastic hatch. In contrast to deck-level mounted antennas are mast-mounted ones shown in Figs. 9–6 and 9–7.

9–5. Radome-enclosed Raytheon antenna/transceiver is mounted directly on the sky-light/hatch. The flexible interconnecting cable to the display console permits hatch being opened when the radar is not in use.

9–6. An EPSCO/Brocks' Seascan radome is mounted on the mizzen mast, well away from lines or sails.

9–7. The Bonzer Nautic-Eye antenna is mounted on a fiberglass baseplate attached to the spreader, with a diagonal bracket from the forward edge of the plate bolted to the mast.

Antennas on Powerboats

Most power cruisers can readily accept the mounting of the antenna on a pilothouse roof, or rigid frame without affecting boat speed or stability. Again, extreme elevation of the antenna does not gain a great deal of range, and accessability of the antenna system for routine maintenance is of greater importance.

If the boat has a flying bridge, there are a number of factors to consider. From an aesthetic or appearance standpoint, antennas are frequently installed on a bracket forward of the bridge, below the windscreen, and on the centerline of the boat. This can result, however, in a blind arc astern of as much as 45-degrees port and starboard.

For a full 360-degree scanning capability, the antenna must be above the flybridge structure. It can be on a short tubular mast elevated above eye level of the helmsman to avoid obstructing his view in all quadrants. Or if there is a signal mast either forward or aft of the flybridge, a bracket can be attached to it to support the antenna.

9–8. A small tubular mast, braced with two diagonal tubes, supports this radome.

9–9. A simple tubular mast supports the 36-inch open scanner and transceiver of the Decca Super-101 on the fly-bridge of this cruiser.

9–10. Smith Industries' Si-Tex scanner mounted directly on pilothouse roof. This 48-inch antenna is used with their Model 22 radar which has a 48-mile range.

9–11. Access to the transceiver, directly below the open scanner, for routine maintenance, is easily achieved due to its location in this installation of a Seavista radar.

9–12. The signal mast on this express cruiser supports the Seascan antenna, without any other modifications or bracing.

Location of Display Console

In many respects the locating of the display for ease of observation is more difficult than the positioning of the antenna. There are a number of points that should be given top priority. First, the presentation on the scope should be aligned parallel with the centerline of the boat. What is shown on the scope should be that which, in clear weather, the helmsman would see looking forward, to either beam or astern. Second, a flat area for charts should be directly adjacent to the display unit, and preferably where both the helmsman and the radar observer can each see them. Third, the mounting of the display should be at an angle that does not require a sharp twist of the neck, when standing or sitting, to look squarely at the scope. Consideration should be given to the use of a seat for the radar observer to use. Fourth, to avoid creating serious deviation to the compass, a "compass safe" distance is usually named by the manufacturer to denote the minimum proximity of the display unit to the steering compass. With some makes it may be as little as 18 inches, but in most cases it is greater than that.

9–13

9-14. When the display unit is mounted close by the steering station it should be far enough away from the compass to avoid creating excessive deviation. In this installation, the Decca 050 appears closer than it should be, and in addition is not aligned parallel with the axis of the boat.

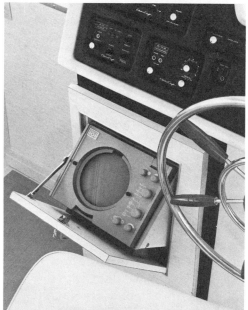

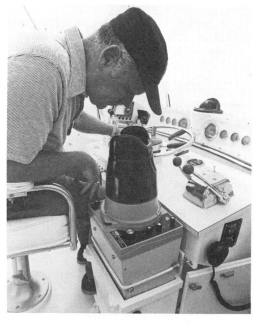

9-15. A hinged panel to support the Decca 101 display is a good means of protecting the unit from weather when the radar is not being used, and when in use the angle of the scope is convenient for both the helmsman and navigator.

9-16. On the flybridge of this cruiser the Raytheon Model 3100 can be dropped down and the hinged box supporting it can be slid under the instrument panel when not being used.

9–17a & b. The Bonzer Nautic-Eye antenna/transceiver assembly is mounted on a tubular mast with diagonal braces on the author's test boat. Small size of the display unit permits its installation where the helmsman or navigator can conveniently view it.

Location of Transceiver

Should the radar be of the three-unit type, with a separate transceiver, other factors need consideration. The distance from the transceiver to the antenna should be as short as possible. Every foot of waveguide is expensive, but of greater importance is the fact that there are losses in transmitting power and receiving capabilities when waveguides must be used. For that reason, a location for the transceiver in a weather-protected, ventilated area, but as near as possible to the antenna, should be selected. This might be on a bulkhead, elevated toward the pilothouse ceiling or within a waterproof (but ventilated) enclosure atop the pilothouse roof. A recently completed moderate-size cruiser has a dummy-stack with the signal mast immediately adjacent to it on which the antenna is mounted. Within the stack is the transceiver, and less than 5 feet of waveguide is required between the transceiver and the antenna. Ventilation of the transceiver is vitally important. While the amount of heat generated within the cabinet is nominal, unless the heat is dissipated through adequate ventilation, serious damage to it can result over a period of time.

Recent developments in flexible cable for use between the transceiver and antenna, in place of rigid waveguide, hold some promise—but generally there is a greater loss of radio energy than with the waveguides.

Location of Power Supply

Many of the more recently developed small craft radars have a solid-state power supply built into either the display console or as an integral part of the radome-enclosed transceiver. There are some models, however, that use a separate unit. In either case, the principal concern is that the power from the boat's batteries be direct and only to the radar. Starting from the distribution panel, where usually there are a multiplicity of fused circuits for running lights, radiotelephone equipment, refrigeration, cabin lights, etc., there should be a separate circuit for radar. In this manner, any load that might be applied to the other circuits and which could cause a temporary drop in line voltage would not affect the radar.

The power supplies of some modern radars have means of compensating for variations in line voltage, and will accept power inputs to their power supplies that may be as much as plus 20 percent to minus 10 percent—without adversely affecting performance of the equipment. Nevertheless, the more nearly the designated input voltage to the radar can be maintained, the better will be the long-term performance of the equipment.

Systems utilizing a separate power supply should have it located as near the main distribution panel, or source of current to it, as possible. This can be in the engine compartment or a well-ventilated locker. If a rotary power supply is used, the axis of it should be fore and aft, rather than thwartship. Bearing life will be shortened if thwartship. Either static or rotary power supplies, if in an engine compartment, should be mounted well above the bilges and in a location where excessive moisture or heat cannot get to them.

Service and Maintenance

Periodically there are certain maintenance procedures necessary with any radar. When positioning the antenna, particularly if the transceiver is a portion of the assembly, you should contemplate the necessity of access to it. If high on a mast this means use of a ladder or a bosun's chair—and the latter is not the most satisfactory perch for a technician with a handful of meters and small tools.

Service of the display console is seldom a problem, except in cases where the console is recessed into a bulkhead. Access panels, or doors, or an easy means of sliding the display console out and into the open should be provided.

Experience of oceangoing commercial vessel operators, covering many years, has shown that maintenance of their radars is substantially reduced if the equipment is left *on* for hours or days at a time—rather than being turned on and off for short periods of time. By leaving the system *on*, components within it are not subject to rapid changes of temperature or electrical load. These frequent changes are more taxing on the components than a continuous load. This same condition exists with television receivers—it is better to leave the unit *on* during an entire evening than to turn it *on and off* at frequent intervals.

Licensing

There are no requirements for the operator of radar on a pleasure boat to have a license. Merchant marine officers, however, must now go through a comprehensive training program to qualify for a radar endorsement to their basic licenses. All radar equipment, though, must be licensed by the Federal Communications Commission, before it can be operated. The reason for this is that a radar transmits radio frequency energy—as does a radiotelephone—and any equipment doing so requires licensing by the FCC.

To get the radar licensed it is necessary to fill out FCC Form 502. In doing so, frequencies for any radiotelephone equipment aboard the boat would also be checked off, even though such equipment was currently licensed. At this time there are no fees involved. When filing for a license through the FCC's Gettysburg, Pennsylvania office, it is necessary to attach a photostat of the current radiotelephone license in order that the original license can be kept aboard the boat while the new license, which includes radar, is being processed. Alternately, the current radiotelephone license and Form 502 which includes both the radiotelephones and radar can be taken to a regional FCC office which will issue an interim permit covering all the equipment.

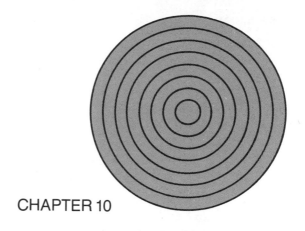

CHAPTER 10

CHOOSING RADAR FOR SMALL CRAFT

The primary purpose of radar aboard a boat is to aid in the prevention of collisions with other craft, or to avoid grounding on a rock, beach, or breakwater. Accidents of these types are at "zero" range. Collisions with either moving or fixed objects occur when the navigator is unable to visually see the danger in time to take evasive action. For this important duty, a radar must be able to show clearly the range and bearing of targets at short distances. Experienced radar operators are unanimous in asserting that the short-range capabilities of a radar are more important than the ability to see targets at 10, 20, or 30 miles, when its use is to avoid collisions or groundings.

The second function of radar is navigation. In many areas of the country, such as Long Island Sound, Puget Sound, Chesapeake Bay, Down East above Boston, the coastlines will be within a relatively few miles. For navigation purposes, a radar with a modest maximum range will be entirely adequate; as it would also be in such other areas as along the Intracoastal Waterway, San Francisco Bay, and Sacramento and Mississippi Rivers. The combination of short-range capabilities as close as 25 yards and a maximum range of 10 to 15 miles can provide both anti-collision safety and navigational assistance for the majority of boat owners.

If offshore cruising is done along the Atlantic and Pacific coasts, or in the Great Lakes, the buyer may want 30- to 48-mile-range equipment for navigation. It should be remembered, however, that if the topography of the coastline normally cruised is low, a radar with long-range capability would be no better than one with 10- to 15-mile range. This is particularly true from the mid-Atlantic states southward to Florida and along the Gulf of Mexico. In contrast, the Pacific Coast has mountains bordering it that are high enough to make a radar with a range capability of 48 miles a great aid in navigating.

On the other hand, a short-range radar used primarily when approaching harbor entrances and within harbors may completely meet the needs of a boatowner. Many boats are well equipped with radio direction finders, Loran or Omega for coastal navigating, and a long-range radar's capabilities for navigation would be redundant. In those cases, a radar that has a maximum range of from 1 to 2 miles would be used for anti-collision purposes when at sea and be entirely adequate when coming into harbors during fog conditions. Such equipment would represent a substantially lower investment, besides being smaller in physical size and lower in weight which would simplify installation of the antenna and display. The foregoing points up the fact that a major factor in selection of a radar is the area most generally cruised and the other electronic navigating equipment aboard the boat. As has been previously stated, to avoid collisions with other craft, breakwaters, and buoys, it is the close-in range that is the most important.

Budget and Operating Considerations

The amount of money to invest in radar and the operational conditions are the principal determining factors in the wise selection of a radar— whether for a tanker or a miniyacht. Higher power and longer ranges are normally needed for larger vessels because of their offshore operations, while equipment with lower power and shorter ranges, costing less than the other equipment, can fully meet the needs of many small craft owners. Considerations that are common to any choice are:

- Experience of the manufacturer in building a wide range of radar equipment for civil and military service.
- Responsibility of the manufacturer as demonstrated by its position in the radar industry.
- Reliability and performance specifications.
- Warranty terms and service facilities.

These are basic value factors in weighing any product, and can be resolved by personal investigation and by discussion with other boat owners who have had experience with radar.

In assessing the operational features of a radar best suited for a particular craft, one's first impression may be a bewildering variety of strange terms and values. One of the purposes of this book is to remove the cloak of mystery, and relate certain design and performance features to those that really matter.

These are the prime points that require consideration:

Cost   Power transmitted
Range performance   Pulse length and frequency
Picture size   Receiver design
Picture quality   Antenna span
Size and weight   Range scales
Power consumption   Controls
Appearance   Installation factors

Range in Relation to Cost

All other factors being the same, increases in maximum range increase the cost of the equipment. Above 12 to 15 miles, for example, it is necessary to have at least two pulse lengths for optimum performance. Above 15 to 18 miles it is desirable to have two or more pulse repetition frequencies in addition to multiple pulse lengths. These require added circuits in the transceiver, and larger antenna systems, both of which add to the cost of the product. Hence the conclusion that if the cruising is mainly in confined waters or reasonably close to a coastline, a quality product of moderate or short range would be the soundest investment. And if the use will be primarily for anticollision or close-quarters navigation, longer range equipment would be an extravagance. In all cases a conservatively rated radar is likely to have a far better performance on small targets, or in confined waters, on its short-range scales than one where the outer limit of its maximum range is the big selling point.

Cost of the radar system is the usual, but not necessarily, the normal method of determining the value of the product. Competition between manufacturers assures this. It will certainly, under ordinary circumstances, determine how much performance one may expect to get. The least costly may not be the best investment, however, and may well not be the cheapest after a year or two of use. To the basic cost of the system, one must add the cost of installation and the cost of maintaining it in good working order. The sum of these in relation to the expected life of the radar represents the true cost.

Short Range Requirements

As noted earlier, a radar capable of as much as 48 miles of range can be extremely helpful in navigating certain coastal areas. But long-range capability, if desired, must be matched with certain factors that insure brilliant performance at short range. It is the last 50 to 100 feet of a radar's short-range capability that can be more important than long range to avoid a disaster.

An extremely short pulse length of from .05 to not more than .1 microseconds is needed to see a target as close as 50 to 75 feet. This must be coupled with a pulse repetition frequency of at least 1500 per second, preferably 2000 to 3000 per second—to paint a bright echo on the scope from a small target.

While an open antenna with a span of 36 to 48 inches will produce a narrower horizontal beam than a radome-enclosed antenna of 30-inch span, on short range scales this is of less significance than when on long ranges. Fig. 10–1 illustrates this point. Hence, for close-quarters navigation or anticollision service, an antenna of as little as 20-inch span enclosed within a radome can be preferable to a wider span open antenna due to its lighter weight and lower power consumption.

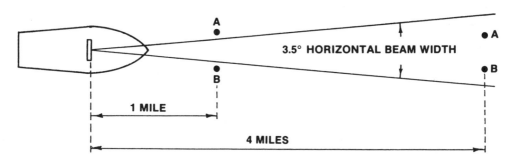

10–1

Picture Size and Quality

The picture size on the scope determines the scale factor for a selected range of operation, and thus determines the ease of viewing. A close parallel may be drawn with a television receiver. To view a large television screen comfortably one must be some distance away from it; conversely a small television screen is better viewed much closer. The radarscope is normally viewed at very short range, and for most small boat operations, a picture tube size of 5 to 6 inches is usually adequate. For those who prefer a larger picture, magnifiers can be used to increase the apparent size of the presentation but at the same time create a certain amount of distortion. Or, if budget and space permit, a larger and more powerful radar may be desirable. As the maximum range performance of a radar increases, so must the size of the picture tube if a reasonable scale factor is to be maintained.

Quality of the picture is vital to the effective use of any radar, and is determined by many factors, such as the antenna, transceiver, picture tube, and the overall design of the equipment. A high-quality picture must be bright, clear, and crisp to facilitate identification of land masses and targets. As in photography, a picture taken by a camera with a poorly ground lens and with limited shutter adjustments will produce a fuzzy print lacking detail and contrast. The same scene taken with a professional camera, using a superior and properly focused lens, and precisely timed shutter, produces a brilliant print. This factor is not easily measured, nor can it be seen when looking at a radar in a dealer's display room, and the buyer must rely upon the reputation and experience of the manufacturer in this respect.

Dimensions and Weight

The size and weight of the radar naturally become more important the smaller the boat on which it will be installed. The growing use of transistors has enabled a considerable reduction in both size and power consumption. Furthermore, when a radar has been especially designed for small craft, the manufacturer usually has taken care to get the radar as compact as is compatible with overall reliability and performance. Assuming use of the best electronic techniques and application of the best possible design methods, further reductions in weight can only be achieved by the use of lighter weight materials and components. Doing so may seriously prejudice the basic ruggedness and reliability of the system. It is important to analyze carefully the operational advantages of a light-weight system in relation to the possible penalties in reliability and performance, which may be sacrificed if the manufacturer has gone to extremes in the direction of lightness and economy.

Engineering, Reliability, and Warranty

Quality of engineering, specifications affecting reliability, and length of warranty are the remaining factors to consider in the selection of a radar. The degree of engineering integrity in design and manufacturing is generally reflected in the reliability of the equipment. Length of the warranty is a clear indication of the manufacturer's confidence in its product to perform day-in and day-out to its specifications.

As long ago as 1957 the Advisory Group on Reliability of Electronic Equipment (AGREE) published a massive document which set out a comprehensive program of procedures to be followed by United States manufacturers of military electronic equipment. They were primarily aimed at increasing reliability, and covered all steps from initial design to mass production.

AGREE standards and procedures for marine radar include: high pressure water spray testing of the antenna; temperature cycling for periods up to 500 hours between 5 and 130 degrees F.; dry-heat cycling up to 158 degrees F. for above deck portions of the radar and to 130 degrees F. for below deck units; low temperature testing to minus 13 degrees F.; shock testing of antennas to as high as 20 G's and to 5 G's for other portions of equipment; vibration tests of 1 G from 0–500 cycles per second and 2 G's for 2-million cycles at all resonant frequencies.

Although radar equipment for military service is generally subjected to more severe strains than on pleasure boats, there is no less a need for the reliability that the AGREE procedures insure when a radar is installed on a small craft. The extent to which a manufacturer follows these procedures should be closely examined before investing in radar equipment, in the interest of efficient operation under the most adverse conditions and to provide maximum safety at all times.

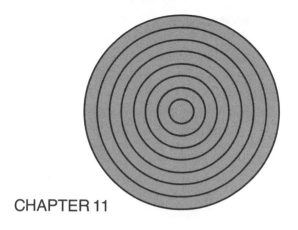

CHAPTER 11

RADAR-ASSISTED COLLISIONS

There is a considerable degree of controversy among mariners and admiralty lawyers over the extent to which radar may have contributed to collisions. Studying some of the classic cases of collisions between vessels that were equipped with radar leads to the conviction that its use gave some navigators a higher degree of confidence and self-assurance than was warranted; and led them to believe that they were operating in a safe manner, when in fact they were not.

The *Andrea Doria/Stockholm* Case

The tragedy that befell these fine ships, less than 200 miles from New York harbor near midnight July 25, 1956 still remains vividly in the minds of those who studied the apparent circumstances leading to the collision. The *Andrea Doria* was nearing the end of its trip from Italy on a course close by Nantucket Lightship; the *Stockholm* was outbound on a course that also passed close by the lightship. These tracks were common to vessels crossing the north Atlantic to or from New York harbor.

For some hours prior to the collision the *Andrea Doria* was in fog, but had reduced speed by only 1 knot from its normal cruising speed of 22.8 knots. Density of the fog varied and was estimated on an average of about 1/2 mile. It was also brought out in the hearings following the collision that at 21.8 knots the *Andrea Doria* required roughly 2 miles to be brought to a stop. The course being steered was 268 degrees when abeam the Nantucket Lightship. The outbound *Stockholm* was just clear of the fog, and proceeding at its normal 18-knot speed on a course of between 087 and 091 degrees. Radars were operating on both ships, but on neither ship was an officer specifically assigned to being radar observer. Instead, the duty was performed intermittently by various officers on the bridge.

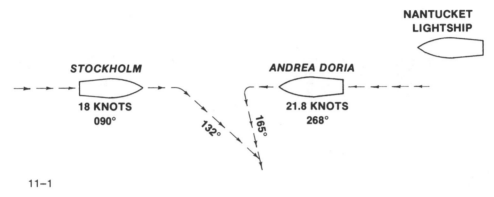

11–1

Testimony at the hearings by the officers of the *Andrea Doria* indicated that they noted a target at 17 miles bearing 4 degrees to starboard. The target was not plotted, nor were plots made at any time between the first radar contact and the collision. On subsequent observations of the radarscope, the target continued approaching the *Andrea Doria*, but (according to the testimony) it was at increasing angles to starboard; this led to their presumption that a starboard-to-starboard side passing could be made. When 3½ miles away from the approaching target, the *Doria* changed course 4 degrees to port—theoretically to widen the passing distance between the ships.

Aboard the *Stockholm* the approaching *Andrea Doria* was observed at 12 miles by the Third Officer on watch, and according to his testimony it was a few degrees off the port bow. Beginning at 10 miles distance, the observations made by the *Stockholm's* officer were plotted, and a port-to-port side passing appeared to be safe, with the closest point of approach estimated at between 1/2 and 1 mile. This presumed, however, that the approaching ship would not change course. To widen the distance between the two ships a 2-degree course change to starboard was made by the *Stockholm*.

Sometime between 3½ to 5 minutes prior to the collision, the *Andrea Doria's* officers saw the running lights of the *Stockholm* off their starboard bow, about 1 mile distant. A hard left turn was ordered, on the belief that a starboard-to-starboard passing could be made.

At about the same time the watch officer on the *Stockholm* saw the lights of the *Andrea Doria*, which until then had been in fog. He did not note that the *Andrea Doria* was starting a left turn. He ordered a hard right turn, anticipating a port-to-port side passing. The result of these opposing maneuvers was that the *Andrea Doria* was struck on her starboard side near the bridge by the *Stockholm*. At the time of collision the *Andrea Doris* was on a course of 165 degrees from the original 265 degrees, and the *Stockholm* was on a course of 132 degrees from the original course of about 090 degrees.

During the hearings a number of points were brought to light that had a strong bearing on the accident. The two ships were not on opposite courses, although nearly so. Neither ship took immediate action of making a bold course change when the other came in view on radar. Lack of plotting by the *Andrea Doria's* officers did not disclose the exact course, speed, or closest point of approach of the *Stockholm*. One of the junior officers on the *Andrea Doria* testified that he would have altered course to starboard, when the echo from the *Stockholmn* was first noted, rather than maintaining its course and later turning to port. He also testified that the course change to starboard would have had to be ordered by the Captain—which, obviously, he did not do. Maintenance of a 21.8-knot speed in restricted visibility by the *Andrea Doria* was in violation of the International Rules of the Road. The climaxing factor was the *Doria's* turn to port when the lights of the *Stockholm* came within sight, and its inability to stop within one-half the distance of visibility.

On the part of the *Stockholm*, the small course change of 2 degrees to the right to widen the port-to-port passing distance was too little a change for it to be apparent to the officers of the *Andrea Doria*—either by radar or visual observation. Had not the *Andrea Doria* made a left turn, the originally plotted closest point of approach of between 1/2 and 1 mile would have been adequate and the 2-degree course change would have merely increased the distance by a small amount. If the course change had been made when the *Andrea Doria* first showed on radar at 12 miles—instead of when they were considerably closer—the distance between them would have been substantially greater when they passed each other.

The Case of *Ships A* and *B*

This collision occurred in the Gulf of Mexico in the early hours of a foggy morning. *Ship A* was on a heading of 284 degrees and cruising at 15 knots when it encountered dense fog. The Master was called to the bridge by the Third Officer, and a half hour later he observed a target at 6 miles range about 22 degrees to starboard. At that time he ordered a course change to port of 14 degrees and shortly afterwards an additional 10 degrees to port, but did not reduce speed. The echoes from the target were not plotted; nor did the change of course of *Ship B* become apparent on the radarscope. Within a few additional minutes the whistle signal of *Ship B* was heard by an officer on the wing of the bridge of *Ship A*, and the engines ordered stopped. This was only seconds before *Ship A* collided with the port after section of *Ship B*.

Little is known of what transpired on the bridge of *Ship B* because most of the crew perished in the fire that followed the collision. An engineroom officer who survived reported that although the bridge had ordered "stand by" to the engineroom, there had been no order to reduce speed from the normal 15 knots, or to stop the engines up to the time of collision. He also reported a marked reduction in engine revolutions shortly before the collision which he related to an abrupt change of course. *Ship B* being hit on its port side indicates that it had turned to starboard in an attempt to avoid *Ship A*; but in doing so put itself directly across the bow of *Ship A*.

In the hearings that followed this accident the paramount criticism was that neither ship had reduced speed when it encountered fog. Disclosure of *Ship B* on *Ship A's* radar at a range of 6 miles, instead of many miles more distant, suggested that the radar was not being put to its full capacity. Had the longer ranges of the radar been used, the course, speed, and closest point of approach of *Ship B* could have been plotted by the officers of *Ship A* and a sufficient change in course made in adequate time to avoid the collision.

The Case of the Tankers at San Francisco

A third element was involved in this case—namely the Harbor Advisory Radar (HAR) experimental installation in operation at the time that the *Oregon Standard* and the *Arizona Standard* collided on January 18, 1971 near the San Francisco Golden Gate Bridge. This installation was shore-based and installed by the Coast Guard to investigate the desirability of harbor advisory systems in this and other U.S. ports. Radar surveillance covered the seaward approach to the Golden Gate, the Bay, and channels leading to it.

Participation in use of the experimental system was purely voluntary. Officers of vessels were not required to monitor VHF channel 18A which was then designated for navigation communications between ships and shore-based stations, but they were encouraged to do so because vessels that volunteered information to the Harbor Advisory Radar Operations Center could be reported by the Center to other vessels if such information was requested. Although the HAR operator saw the impending collision, the lack of authority and inability to communicate with the navigator of *Oregon Standard* and lack of mandatory VHF bridge-to-bridge communication, contributed to the calamity.

What preceded the collision is worthy of note: the *Oregon Standard* left its dock at Richmond shortly after midnight outbound in dense fog. It reported to HAR on VHF channel 18A that it was underway, at about 0049 hours, then switched the VHF radio to a company frequency and later to channel 6 instead of 16. Two radars were in operation, but both on short range scales. Meanwhile the *Arizona Standard* was inbound to San Francisco and heard the exchange of information on VHF channel 18A between the *Oregon Standard* and HAR while still seaward of the Golden Gate Bridge. Its radar was on a medium-range scale.

According to testimony at the hearings into causes of the collision, the *Oregon Standard* was observed by the *Arizona Standard* on its radar at about 6 miles distance and to the east of the Golden Gate Bridge. For reasons that were never explained, the echo from the *Oregon Standard* became lost on the radarscope of the *Arizona Standard*—possibly because the echo from the bridge lay across the courses being steered by the two tankers.

Until shortly before coming to the bridge, the outbound *Oregon Standard* had been operating at 10.5 knots, and the inbound *Arizona Standard* at 11.4 knots. Speed on the *Oregon* was reduced to 7 knots shortly before reaching the bridge. Both tankers continued these respective speeds until the collision.

A number of radio communications—mainly related to the movement outbound of the *Oregon*—were exchanged between HAR and the inbound *Arizona* on channel 18A. Attempts were made a number of times by the *Arizona* and HAR to raise the *Oregon* on channels 16 and 18A, but without success. It was determined in the hearings that the *Oregon's* VHF was inadvertently tuned to channel 6, instead of 16 or 18A.

The accident investigation further disclosed that the outbound *Ore-*

gon had its two radars on short ranges—one on the 3-mile range, the other on 5 miles but later shortened to 1½ miles when approaching the bridge. On neither scale was the inbound *Arizona* within range until the last few minutes before the collision. Had one of the radars on the *Oregon* been on a longer range scale—such as 10 miles—presence and position of the *Arizona* would have been evident to the *Oregon*.

According to the investigation, another contributing factor to the collision was that the outbound *Oregon* was too far south of the normal outbound traffic lane; and the *Arizona* too far north, putting both ships virtually in the center of the channel. This is in contradiction to the Rules of the Road which specify that a vessel must stay on the right side of a channel. The final contributing factor occurred when the inbound *Arizona* was within visual range of the *Oregon* and made a turn to port instead of to starboard.

Final conclusions from the hearings placed blame on the Masters of each tanker for excessive speed, for failure to stay on the right sides of the channel, for improper usage of information available from the radar equipment aboard the vessels, and for failure to properly use VHF radio communications channels that would have permitted exchange of information as to position and intentions, or to utilize the information available from the Harbor Advisory Radar Center.

In the November 1971 issue of *Proceedings* of the Marine Safety Council (Department of Transportation, U.S. Coast Guard) is a detailed report of the causes and results of this collision. The report included the radarscope photographs, Fig. 11-2, a, b, and c, as well as chart reproductions showing the tracks of the two tankers up to the time of collision.

The chances of a collision of this type occurring again have been substantially reduced within the last few years. The Bridge-to-Bridge Communications Act that became effective January 1, 1973 requires all vessels of 300 gross tons or more, passenger carrying vessels of 100 tons or more, dredges and tugs of 26 feet or more engaged in towing to maintain a continuous radio-watch on VHF channel 13. It is specifically required that channel 13 be used for no other purpose than exchange of navigational information between ships, or for communication with shore-based stations used for vessel traffic control. The experimental Harbor Advisory Radar system, in use at San Francisco when the *Oregon* and *Arizona* collided, is now a part of the Coast Guard's Vessel Traffic Systems. No longer is VHF channel 18A used on a voluntary basis; instead, at San Francisco and all other harbors where

11–2a

11–2c

11–2a, b, & c. Radarscope photos of the Coast Guard's Harbor Advisory Radar at San Francisco, showing in Fig. 11–2a, the outbound *Oregon Standard* indicated by arrow at 0136:08 to the east of Golden Gate bridge; in Fig. 11–2b the *Oregon Standard* passing under the bridge, and the inbound *Arizona Standard* approximately 900 yards west-southwest of the bridge at 0139:02; and in Fig. 11–2c the echoes from both tankers merged in a position approximately 150 yards west-south-west of the center of the bridge at 0141:58 when the two ships collided. (U.S. Coast Guard Official Photos)

11–2b

Vessel Traffic Systems are in operation, channel 13 is mandatory and (in addition to a vessel also monitoring channel 16) used for both ship/ship and ship/shore communications involving vessel movement.

In Retrospect

In looking back on these accidents between radar-equipped vessels, manned by professional mariners, it appears to be more than mere coincidence that the contributing factors were primarily the same in each of these cases. Excessive speed during times of limited visibility and inability to stop the vessels within one-half the range of visibility were true in each case. Had the vessels been operated at "moderate" speed, it is probable that the collisions could have been avoided.

The second common denominator was the failure to fully utilize the information available from the radars. This involves periodic scanning of the ocean on the medium- and long-range scales alternately with use of the short ranges, instead of keeping the range set to a short distance which does not allow sufficient time between detection of a target and the taking of evasive action. Coupled with this factor was the failure to plot course, speed, and closest point of approach of the approaching ship, and based on the plot the taking of immediate and substantial evasive action. The final factor was the turning to port by one ship, in each of the three collisions, instead of making all turns to starboard.

As was stated a number of times earlier in this book—*radar is an aid to navigation*, it *cannot* do the navigating. It can present valuable information to the navigator that, if properly interpreted, can prevent collisions with other boats or objects; it can greatly assist in solving navigation problems; but it is only an electronic device that is intended for the use of man, not as a substitute for him.

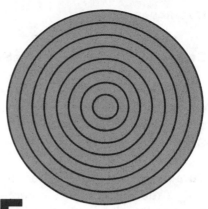

CHAPTER 12

GLOSSARY OF RADAR TERMS

Aerial—an alternate word for antenna or scanner

Afterglow—the continuing glow of an echo on the cathode ray tube after the sweep trace has painted an object or landmark on the tube

Antenna—the part of a radar system from which radio waves are transmitted and their echoes received

Anti-clutter Control—a control of the receiver circuit to reduce intensity of echoes created by wave action or sea-return

Aperture—the effective radiating span of an antenna, usually expressed as the horizontal width of the scanner

Attenuation—weakening of radio wave power due to absorption or scattering

Azimuth—the bearing circle surrounding the perimeter of the radarscope, marked in degrees, to measure relative bearing of an echo

Beam Width—angular width, horizontal and vertical, of the radiated radio waves from the antenna or scanner

Bearing—the direction to a target either relative to the bow, or magnetic, as measured on the azimuth

Cathode Ray Tube—the display tube on which objects are painted, around which is an azimuth scale to measure bearing of echoes created by targets

Clutter—echoes created by waves, rain, or snow which appear at random on the radarscope and which tend to obscure weak echoes from small targets

Co-axial Cable—a flexible cable to conduct radio frequency energy between the transceiver of a radar system and the display console, with a metal braid shield between the conductors and the outside insulation

Compass Bearing—the magnetic bearing of an object or target with respect to the boat from which the bearing is taken

Crystal—a synthetic-quartz element to control the frequency of radio waves

Cursor—a manually rotated transparent disc over the display tube with a radial line from the center to the azimuth to determine the relative bearing of an object shown as an echo on the radarscope

Definition—a measure of the quality of the presentation of echoes on the radarscope; sharpness of detail

Differentiation—a circuit in the receiver designed to reduce clutter from rain, snow, and waves; improves definition at short ranges

Discrimination—the ability of a radar to separate the echoes from two targets that are close in range or bearing

Duct—strata created by atmospheric conditions that tends to confine radio waves within the strata; ducting extends the range beyond normal ranges.

Echo—the reflected radio waves from a target; the spot or "blip" shown on the radarscope created by the reflected radio waves

Electron Gun—that portion of the cathode ray tube which emits a stream of electrons to the fluorescent-coated inner face of the tube, the intensity of which is increased to paint a spot when an echo is reflected from an object

Fluorescence—light emitted from the cathode ray tube as the result of a high charge of electrons striking the tube face; *See* Afterglow and Persistence

Gain—as applied to the radar receiver, it is adjustable to control the echo power as displayed on the scope; corresponds to the volume control on a radio

GCA (Ground Control Approach)—system used at airports for aircraft traffic control

Heading—the direction in which a boat is moving; as used in this book: the magnetic course being steered

Heading Flasher—the electronically generated line on the radarscope indicating the bow of the boat; alternate term for Heading Marker

Interference—intermittent and random paints on the radarscope generally created by signals on the same frequency by radars aboard other boats, or from shore-based radars

Ionosphere—a layer of heavily ionized molecules in the outer part of the earth's atmosphere, far beyond the stratosphere

Line-of-sight—distance to the horizon; range at which objects can be visually perceived.

Magnetic Bearing—the bearing shown by the compass of an object with respect to the boat from which the bearing is taken

Microsecond—one millionth of one second

Microwaves—very short radio waves of a length in which radar operates

Mixer—the portion of a radar receiver which changes the frequency of the received radio waves to a frequency that permits amplification of the weakest signals to a level that will show on the scope

Object—a boat, buoy, landmark, or other radar reflective mass; *See* Target

Persistence—continuation of the glow of an echo on the radarscope after it has been painted by the discharge from the electron gun

Plan Position Indicator—the display as shown on a modern radarscope; chartlike presentation of shorelines, boats, or buoys on the scope

Power Consumption—amount of electric power required to operate a radar; usually expressed in watts

Pulse—emissions of radio frequency energy; bursts of energy, as contrasted to a continuous emission of energy

Pulse Length—length in time, measured in microseconds, that pulses of radio energy are transmitted

Radar—contraction of Radio Detection and Ranging

Radarscope—alternate term descriptive of display or cathode ray tube or tube or scope, on which the echoes from landmarks, objects, or targets are shown

Radome—a fiberglass housing used to enclose the rotating scanner and other elements of the antenna system

Range Ring—concentric circles of electronically generated light on the radarscope to give visual indication of distance in nautical miles to echoes created by objects; used to measure distance. Sometimes fixed, painted, or scribed lines for the same purpose

RaRef—abbreviation used on U.S. charts to indicate that the buoy is equipped with a radar reflector to increase the range at which it is radar-conspicuous.

Relative Bearing—the bearing of an object in respect to the boat from which the bearing is taken; the number of degrees to port or starboard of the bow

Relative Motion—motion of other boat in relation to radar-equipped boat

Reflector—a safety device that can be hoisted on a boat's mast consisting of three planes of metal, each plane mutually at right angles to each other to produce a stronger radar echo

Repetition Frequency—number of times per second that pulses of radio energy are transmitted; alternate term to Pulse Repetition Rate

Resolution—the measure of a radar's ability to separate two objects that are close to each other in range or bearing, and to show them as two distinct echoes on the scope; alternate to Discrimination

Sea-Return—the random echoes created by cresting waves, generally in the quadrant of the scope from which the wind is blowing

Scanner—the rotating element of the antenna system that emits radio waves and picks up the returning echoes; alternate to Aerial or Antenna

Scope—the fluorescent face of the cathode ray tube

Sensitivity—the measure of a radar receiver's ability to detect weak signals

Signal—reflected radio waves or echoes

Slotted Waveguide—the portion of the scanner assembly in which apertures have been cut for emission of radio waves and collection of returning echoes or signals

Solid State—equipment that uses transistors or other solid-state devices such as integrated circuits in place of vacuum tubes

Span—horizontal width of the antenna or scanner measured between the two extremes of its aperture.

Sweep—radial line displayed on the cathode ray tube, synchronized with the rotation of the scanner, on which echoes will appear as bright spots in relation to their distance

Target—term used to denote a boat, buoy, island, headland, or other object that is radar-conspicuous and produces an echo on the radarscope

Trace—alternate term to sweep

Transceiver—contraction of transmitter and receiver

Transducer—an electro-mechanical device to convert electrical energy into mechanical energy, and vice versa; in depth sounders it converts electronic impulses into sound impulses, or vice versa

Transistor—a device to amplify, generate, rectify, or convert electrical energy more efficiently than vacuum tubes

Waveguide—a hollow metal duct, formed to precise dimensions, through which radio waves travel between the transceiver and scanner

BIBLIOGRAPHY

There are a number of authoritative books on marine radar theory, design, construction, and usage which those interested in the more technical aspects of radar may find instructive. Most are written for use by engineers or merchant marine officers, and the titles indicate, to a degree, the content of the books.

Brown, Ernest B. *Radar Navigation Manual.* H. O. Publication No. 1310. Washington, D.C.: Defense Mapping Agency Hydrographic Center.

Burger, W. *Radar Observer's Handbook.* 52 Darnley St., Glasgow, Scotland. Brown, Son & Fergueson, Ltd., 1961.

French, John. *Small Craft Radar.* New York: Van Nostrand Reinhold, Co., 1977.

Moss, Captain W. D. *Radar Watchkeeping: For the Professional Seaman*, 2nd Ed. London: *The Maritime Press Ltd.* 1973.

Oudet, Captain L. *Radar and Collision.* Princeton: D. Van Nostrand Company, 1960.

Slack, Captain Robert M., and Whittle, Patrick. *Marine Radar Accidents.* San Francisco: Bancroft-Whitney Co.

Sonnenberg, G. J. *Radar and Electronic Navigation*, 4th Ed. New York: Van Nostrand Reinhold Co., 1970.

Wylie, Captain F. J. *Choosing and Using Ship's Radar.* New York: American Elsevier Publishing Co., Inc., 1970.

———. editor. *Use of Radar at Sea.* New York: American Elsevier Publishing Co., Inc., 1968.

SUPPLIERS

A partial list of U.S. manufacturers and suppliers of radar equipment for small craft includes the following firms:

Bonzer, Inc. (Nautic-Eye)
90th and Cody, Overland Park, Kansas 66214

EPSCO/Brocks, Division of Epsco, Inc. (Seascan and Seaveyor)
12 Blanchard Road, Burlington, Massachusetts 01803

ITT Decca Marine
Palm Coast, Florida 32037

Konel Corp. (Konel/Furuno)
271 Harbor Way, South San Francisco, California 94080

Raytheon Company
676 Island Pond Road, Manchester, New Hampshire 03103

Ray Jefferson Co.
Main & Cotton Streets, Philadelphia, Pennsylvania 19127

Smiths Industries, Inc. (Si-Tex)
St. Petersburg/Clearwater Airport, Clearwater, Florida 33518

INDEX